Hurricanes

Hurricanes
Their Nature and
Impacts on Society

Roger A. Pielke, Jr

National Center for Atmospheric Research

and

Roger A. Pielke, Sr

Colorado State University

JOHN WILEY & SONS

Chichester • New York • Weinheim • Brisbane • Singapore • Toronto

Other Wiley Editorial Offices

John Wiley & Sons, Inc., 605 Third Avenue,
New York, NY 10158-0012, USA

WILEY-VCH Verlag GmbH, VCH Verlagsgesellschaft mbH,
Pappelallee 3, D-69469 Weinheim, Germany

Jacaranda Wiley Ltd, 33 Park Road, Milton,
Queensland 4064, Australia

John Wiley & Sons (Canada) Ltd, 22 Worcester Road,
Rexdale, Ontario M9W 1L1, Canada

John Wiley & Sons (Asia) Pte Ltd, 2 Clementi Loop #02-01,
Jin Xing Distripark, Singapore 129809

Library of Congress Cataloging-in-Publication Data

Pielke, Roger A., 1968–
    Hurricanes : their nature and impacts on society / Roger A.
Pielke, Jr. and Roger A. Pielke, Sr.
        p. cm.
    Includes bibliographical references and index.
    ISBN 0-471-97354-8
    1. Hurricanes—Government policy—United States. 2. Hurricanes—
Social aspects—United States. I. Pielke, Roger A. II. Title.
QC945.P64 1997
363.34'922'0973—dc21   97–14116
                                                        CIP

British Library Cataloguing in Publication Data

A catalogue record for this book is available from the British Library

ISBN 0-471-97354-8

Typeset in 10/12pt Times from authors' disks by Mayhew Typesetting, Rhayader, Powys
Printed and bound in Great Britain by Biddles Ltd, Guildford and King's Lynn

This book is printed on acid-free paper responsibly manufactured from sustainable forestation, for which at least two trees are planted for each one used for paper production.

Contents

Dedication ix
Foreword xi
Preface xv

CHAPTER 1 **Introduction: Science, Policy, and Hurricanes** . . . 1
  1.1 The Hurricane: "A Melancholy Event" 1
  1.2 "We Need Help": Hurricane Andrew in South Florida,
      August 1992 2
      1.2.1 *Forecast* 3
      1.2.2 *Impact* 7
      1.2.3 *Response* 10
  1.3 Defining the Problem 14
      1.3.1 *The hurricane as an extreme meteorological event* . . 15
      1.3.2 *Hurricanes in North American history* 16
      1.3.3 *Ten notable hurricanes of the past century* 19
      1.3.4 *Extreme weather events* 25
      1.3.5 *Science in service to society* 26
      1.3.6 *The challenge: toward a more usable science* . . . 30

CHAPTER 2 **The US Hurricane Problem** 31
  2.1 Reframing the US Hurricane Problem 31
      2.1.1 *The challenge of problem definition* 31
      2.1.2 *The conventional framing of the US hurricane*
           problem 35
      2.1.3 *Societal vulnerability as an alternative framing of*
           the hurricane problem 37
  2.2 Vulnerability to hurricanes 38
      2.2.1 *Hurricane incidence* 39
      2.2.2 *Hurricane incidence and climate change* 47
      2.2.3 *Exposure to hurricanes* 49

2.3   Assessment of Vulnerability to Hurricanes.   59
      2.3.1 *Tropical cyclone risk assessment*   59
      2.3.2 *Societal vulnerability to tropical cyclones: a*
             framework for assessment 64
      2.3.3 *Incidence assessment*   65
      2.3.4 *Exposure assessment*   65

CHAPTER 3   **Tropical Cyclones on Planet Earth** 68
3.1   Life of a Hurricane 68
      3.1.1 *Birth and growth* 68
      3.1.2 *Maturity* 72
      3.1.3 *Decay.* 79
      3.1.4 *Criteria for development and intensification of a*
             tropical cyclone 82
3.2   Special Cases of Development and Intensification 83
3.3   Geographic and Seasonal Distribution.   85
      3.3.1 *Origin*   85
      3.3.2 *Movement.*   85
      3.3.3 *Tropical cyclones in the Atlantic Ocean Basin* .   .   . 87

CHAPTER 4   **Hurricane Forecasts**   92
4.1   Tropical Cyclone Movement 92
4.2   External Flow: The Steering Current 92
4.3   Interaction of the Steering Current and the
      Hurricane   99
4.4   Internal Flow.   102
4.5   Tropical Cyclone Track, Intensity, and Seasonal
      Forecasting 102
      4.5.1 *Tropical cyclone track predictions*   102
      4.5.2 *Tropical cyclone intensity change predictions*   .   . 106
      4.5.3 *Seasonal predictions of tropical cyclone activity* .   .   . 109
      4.5.4 *Attempts at tropical cyclone modification*   .   .   . 111
      4.5.5 *Value to society of forecasts*   113

CHAPTER 5   **Hurricane Impacts.**   118
5.1   Ocean Impacts 118
5.2   Land Impacts at the Coast and a Short Distance
      Inland   119
      5.2.1 *Storm surge* 120
      5.2.2 *Storm surge analysis*   121

5.2.3   *Winds.* 122
5.2.4   *Rainfall* 125
5.2.5   *Tornadoes.* 125
5.2.6   *Inland impacts.* 127
5.3   Societal Impacts 131
5.3.1   *Hurricane impacts on society* 131
5.3.2   *The challenge of estimating damages* 134
5.3.3   *Summary* 137

CHAPTER 6   **Societal Responses.** 139
6.1   Understanding Societal Responses to Weather Events . . . 139
6.2   Long-Term Social and Decision Processes 141
6.2.1   *Preparing for evacuation* 141
6.2.2   *Preparing for impacts* 142
6.2.3   *Preparing for recovery* 144
6.3   Short-Term Decision Processes 148
6.3.1   *Forecast: the art and science of hurricane track
        prediction* 148
6.3.2   *Impact: surviving the storm* 152
6.3.3   *Response: recovery and restoration* 153
6.4   Conclusion: Preparedness Assessment 154

CHAPTER 7   **Hurricane Andrew: Forecast, Impact, Response** . . . 156
7.1   Introduction 156
7.2   Forecast 156
7.2.1   *Hurricane track and intensity* 156
7.2.2   *Evacuation* 158
7.3   Impact 163
7.3.1   *Direct damages from Hurricane Andrew* . . . 163
7.3.2   *Building codes: construction, implementation,
        enforcement, and compliance* 170
7.3.3   *Insurance* 176
7.4   Response. 177
7.4.1   *Recovery* 177
7.4.2   *Restoration* 179

CHAPTER 8   **Tropical Cyclone Fundamentals** 181
8.1   From Knowledge to Action 181
8.2   Ten Important Lessons of Hurricanes 182
8.3   Last Words 191

APPENDIX A Additional Reading 194

APPENDIX B Economic and Casualty Data for the United
            States 198

APPENDIX C Selected Data on Tropical Storm and Hurricane
            Incidence in the Atlantic Ocean Basin. 204

APPENDIX D Selected Data and Names of Tropical Cyclones
            in Ocean Basins Around the World 210

APPENDIX E Guide for Local Hurricane Decision-Makers . . . 232

APPENDIX F Units 256

References 258

Index 276

Dedication

We dedicate this book to the planners, engineers, builders, forecasters, emergency managers, relief workers, and other public and private officials and volunteers who have committed their lives to preparing for and responding to hurricane impacts on society.

Foreword

Before the mid-1960s, studies of the societal impact of hurricanes in the United States were founded mainly on the statistics of mortality and economic losses. These statistics, showing a significant steady reduction in the ratio of lives lost to property damage, were widely (and properly) used as evidence of increasing effectiveness of hurricane warning services. Some researchers, however, found these figures more troublesome than encouraging, in view of the rapidly increasing amount of property at risk along southern US coastal areas followed inevitably by expansion in numbers of lives in harm's way.

During the 1940s and 1950s, concern grew in Congress over the progressive increases in the number of hurricanes affecting the United States each year. This concern reached boiling point in 1954 when three major hurricanes, Carol, Edna, and Hazel, ravaged coastal areas from the Carolinas to Maine. The response was a recurring appropriation, beginning in 1955, of considerable funds for hurricane research and new technologies to deal with the perception of an increasing annual threat from hurricanes. This infusion of new funds for programs to minimize losses of life and property, led social and physical scientists and engineers individually to accelerate their efforts in search of more effective means and options for disaster mitigation. It was a single catastrophic event, however – Hurricane Camille of 1969 – which brought these separate efforts and disciplines together in recognition of the multi-faceted nature of the problem.

Hurricane Camille, which ravaged the coastlines of Louisiana, Mississippi, and Alabama early on the morning of 18 August 1969, brought a new dimension for considering the destructive potential of a major hurricane, and the variety of societal responses involved. Camille brought a storm surge of nearly 25 feet, the highest ever recorded in the Western Hemisphere, the second lowest pressure on record in the Western Hemisphere at that time, maximum winds of record hurricane strength, record property losses of $1.4 billion, and the loss of 256 lives. Finally, after moving inland more than a thousand miles, it deposited maximum point rainfall amounts – in excess of 27 inches in Virginia – before moving out into the open Atlantic and regaining hurricane wind strength. As a consequence of warnings, more than 75 000 were evacuated inland from exposed coastal locations.

The lessons of Camille supplied a bridge of understanding between the physical and social sciences and engineering. It was a convincing example of the need to view the tasks of mitigation in holistic perspective as a multi-disciplinary problem of optimizing not only science and technology to produce warnings and preparedness methods, but also concomitantly the analysis and societal responses in the face of a hurricane threat.

Perhaps the most important catalyst in the ongoing efforts to forge public policies, methods, and practices to minimize potential losses in disasters was the establishment at the University of Colorado in Boulder of a Natural Hazards Center, under the leadership of Gilbert F. White. That institute not only conducted its own programs of research on this subject, but sponsored annual conferences of leading researchers in the fields of physical and social science, and policy-level officials of state and federal government to review the progress of research and probe the more difficult problems encountered in each area, including policy planning. This book, primarily devoted to the definition and analysis of the problems to be faced in pursuing realistic goals for reducing potential losses from hurricanes, is rooted in the knowledge gained from, or stimulated across the nation by, this Center.

This book, which begins with a riveting description of the hurricane disaster scene from eyewitness and authoritative accounts, addresses the question: "What is the social importance of funds poured into hurricane research and supporting technology?" Today, Congress and funding agencies are demanding clearer evidence of benefits to be expected from the investment of further funds on hurricane mitigation research. It is argued that, even though annual losses of life have diminished dramatically, economic losses have nevertheless skyrocketed. Moreover, from overall societal considerations, we are more vulnerable than ever before – more people are at risk and incredibly more property is in harm's way. The conclusion is drawn that the present hurricane policy of "minimizing loss of lives and property" is not a particularly useful way of framing the goal of a US hurricane policy. It is suggested that a better approach would be to define the goals as a function of societal vulnerability. From this baseline, the authors supply an extensive description of the physical nature of the hurricane and of its prediction, followed by an analysis of the societal problems of response to warnings and to the administration of miti-gation measures – preparedness, relocation, and recovery. The challenge posed is to redefine the policy goals of the US hurricane program in terms of minimizing the potential societal vulnerability considering the realistic limita-tions of warning skills and of societal responses. While this treatise does not purport to supply explicit answers, it does provide a framework for planning and seeking answers to the questions that are addressed.

This book is the unique product of a father–son effort, both well equipped by training and experience. Roger Pielke Sr spent his maturing years as a research meteorologist and theoretician in the milieu of NOAA's National Hurricane Center. Always the kind of theoretician eager to understand and become involved in the application of his science, he was usually to be found "deep in the front line trenches" during the battle of wits with hurricanes at that center. From his auspicious start at Miami and later at the University of Virginia and Colorado State University, he went on to become a leading authority in modeling smaller-scale weather circulations and disturbances, and author of several textbooks, including one on hurricanes. Roger Pielke Jr, raised in the environment of hurricane concern at Miami, channeled his interests into social and policy aspects of the science/society relation. Today, he continues his research on public policy in regard to weather and climate at the Environmental and Societal Impacts Group at the National Center for Atmospheric Research. This father–son effort has profited from their respective complementary concerns, backgrounds, and experiences in addressing a complex problem of importance to us all. Their book should provide an excellent starting point and guide in the development of public policies and in planning the support and funding of public programs for disaster mitigation, including preparedness and recovery.

R.H. Simpson

Preface

The hurricane is one of nature's most intense phenomena and one of the coastal resident's greatest fears. Through coastal development and growth, the United States has become more vulnerable to the impacts of hurricanes than at any time in the recent past. This claim is supported by damages averaging more than $6 billion annually over the period 1989–1995 (adjusted to constant 1995 dollars), as compared to $1.6 billion annually for the period 1950–1989. In spite of the remarkable scientific and technical advances made this century, it is due in significant part to luck that there has not been a large loss of life in the United States due to a hurricane's landfall – hurricanes have not recently struck our most vulnerable places. As society's vulnerability continues to grow, hurricane impacts will become a more important policy problem.

One of our main purposes in writing this book is to explore and define the problem posed to society by hurricanes in the hope that with a better understanding of the problem, decision-makers will be in a better position to formulate and implement effective policies. Our main goal is to provide a foundational portrait of the interrelated environmental and societal aspects of the hurricane problem, hoping that others with authority and responsibility for decision might better understand the nature and impacts of hurricanes. We will consider the book a success if it stimulates debate, discussion, and further reflection on the hurricane as a societal problem and if this dialogue has a positive influence in the policy process.

Throughout the book, we use the concept of "vulnerability" to integrate the environmental and societal aspects of the hurricane problem. In using the term vulnerability, we mean to define explicitly the hurricane problem as a joint function of extreme environmental phenomena (i.e. natural hazards such as wind, flooding, etc.) and of human exposure to those phenomena. In the past, much of society's understanding of the relation of environmental extremes and human impacts was grounded in a framework of "nature-causing-disaster". Today, there is a much greater appreciation throughout the community of natural hazards scholars and practitioners that, in addition to nature, people are in large part responsible for disasters through choices that they make (or do not make) and decisions that others make for them. We seek to bring this integrated perspective to the case of hurricanes.

One of the fundamental assumptions that we bring to this book is that improving policies in response to the hurricane problem first requires that the problem be well understood and defined. Our sense, as discussed in the book, is that in many respects the problem posed by hurricanes to society is not well defined. In assuming that developing more effective solutions depends upon first understanding the problem being faced, we do not seek to imply that we can either fully understand the problem or can develop solutions without the hard-earned knowledge gained from practical experience. In fact, we do not believe that there is *a* solution to the hurricane problem. Rather, there will be many solutions in many different places. What works for Miami will not necessarily work for New Orleans, or for Kingston, Jamaica for that matter. Most "solutions" will have to be field tested and evaluated, with successful policies kept, refined, and extended, and failed policies documented and terminated. We intend this book as a starting point – a large-scale map of the policy problem that leaves the invention, selection, and implementation of specific policy alternatives to those more capable to the task. We hope that with the rough map that we provide, the individual, community, or national decision-maker will be more prepared to fill in details based on the particular context that they face.

A second, broader theme of this book is the relationship between science and society. In recent years, United States science policy has witnessed important changes, among them calls by policy-makers and the public that federally funded research show more direct usefulness and value with respect to society's needs (of course, demands for increased accountability of government go well beyond science). We feel strongly that accountability of science is healthy for both the sponsors and conductors of research. Yet one characteristic of the changing environment of science policy is that the scientific community lacks effective methods and experience to demonstrate its value to society.

We feel that the atmospheric sciences, and the area of hurricane research specifically, is particularly well suited to demonstrate the value of its research to society. Of course, research is of little use or value unless its results can be incorporated into actual decisions. (We also recognize the intrinsic value of knowledge.) To be made useful, scientific research must be integrated with the needs of people seeking to address problems or opportunities that they face. Consequently, we believe that a problem-oriented approach will set the stage for identification of those aspects of decisions that might be improved with the results of research, those decision processes that might be altered to take advantage of information produced through research, and importantly – to identify those aspects of the hurricane problem that are largely independent of scientific research. We believe that by beginning with the problems posed to society by hurricanes, it will be relatively straightforward to identify where current and future research will likely have the greatest payoffs. Many of these payoffs will likely come in decision contexts and situations different than expected.

In order to assist the reader who requires more information, detail, or a more sophisticated discussion of the various topics that we cover, we have provided a list of additional readings in Appendix A. This compilation of articles, books, and WWW sites provides substantial insight into the nature and impacts of hurricanes that goes well beyond what is found here. Our approach is grounded in the different perspectives that we bring to this issue. Roger Sr was trained as a meteorologist at Penn State University and had the fortunate opportunity to work from 1971 to 1974 for Joanne Simpson at the US National Oceanic and Atmospheric Administration's Experimental Meteorology Lab (EML) in Miami, which was co-located with the National Hurricane Center, then with Robert H. Simpson as Director. While at EML, Roger Sr had the opportunity to know and learn from a number of hurricane forecasters and tropical meteorology experts including Rick Anthes, Pete Black, Bob Burpee, Gil Clark, Marina Estoque, Cecil Gentry, Cecilia Griffith, Neil Frank, Harry Hawkins, Paul Herbert, Ron Holle, Brian Jarvinen, Bob Jones, Miles Lawrence, Banner Miller, Charlie Newmann, Jose Fernandez-Partagas, Joe Pelisser, Stan Rosenthal, Bob Sheets, Arnold Sugg, Jim Trout, Victor Wiggert, and Bill Woodley. In 1974, Roger Sr, along with Bob and Joanne Simpson, began teaching at the University of Virginia where he was fortunate to interact in tropical meteorology with Dave Emmitt, Bruce Hayden, Mike Garstang, and Pat Michaels. In 1981, Roger Sr joined the faculty at Colorado State University (CSU) where he has profited from the knowledge and expertise in this subject area of Bill Cotton, Steve Cox, Bill Gray, Dick Johnson, John Knaff, Chris Landsea, Mike Montgomery, John Scheaffer, Wayne Schubert, Tom Vonder Haar, and Ray Zehr.

Roger Jr was trained as a political scientist at the University of Colorado where he focused on science policy under the guidance of Ron Brunner, Susan Clarke, and Sam Fitch. At Colorado, Roger Jr was trained in the tradition of the policy sciences, a distinctive tradition in the policy movement, characterized by an approach to the study of policy that is contextual, problem-oriented, and multi-method. In 1991, Roger Jr had an opportunity to work for the House Science Committee in Washington, DC, then under the leadership of Representative George Brown. The Chief of Staff of that Committee, Rad Byerly, has taught Roger Jr much about the relation of science and policy-making, both at the Committee and since. In 1994, Roger Jr joined the staff of the Environmental and Societal Impacts Group (ESIG) at the National Center for Atmospheric Research (NCAR) where he has benefited from a close working relation with Mickey Glantz, a world leader in the study of the relation of climate and society.

The idea for this book has its origin in a post-doc position funded by the National Science Foundation that Roger Jr spent at ESIG studying societal responses to extreme weather events. A National Oceanographic and Atmospheric Administration (NOAA) conference on hurricanes organized by Henry Diaz and Roger Pulwarty provided the first opportunity for us to collaborate on the subject of hurricanes, bringing together some of Roger Sr's earlier work with Roger Jr's post-doc research. With so much discussion in recent years about the need for social scientists to work more closely with physical scientists, we figured that our collaboration ought to be a best-case scenario and that we should be able to overcome the traditional obstacles to multidisciplinary collaboration. (We also figured that if a father–son team could not overcome those obstacles, then the goal of increased social/physical science interaction ought to be rethought!)

Over the past three years, we received guidance and help, both small and large, from many people with expertise in hurricanes, weather, and natural hazards. These people kindly provided much research assistance and many valuable suggestions, comments, and answers to our questions. These include Emery Boose, Bob Burpee, Jack Cermak, Leighton Cochran, Bill Cotton, Mark DeMaria, Steve Doig, Neil Dorst, Gervaise Dupree, Steve Dickson, Joe Eastman, Leslie Forehand, Joe Elms, Stan Goldenberg, Greg Holland, Susan Howard, Karen Gahagan, Mickey Glantz, Bill Gray, Chip Guard, Bill Hooke, Jerry Jarrell, Brian Jarvinen, John Knaff, Jim Kossin, Steve Lord, Walt Lyons, Brian Maher, Kishor Mehta, Bob Meroni, Dennis Mileti, Dave Morton, Mary Fran Myers, Steve Nelson, Charlie Neumann, Mel Nicholls, Fid Norton, Gary Padgett, John Peabody, Jon Peterka, Bob Plott, Mark Powell, Jim Purdom, Jessica Rapp, Lory Reyes, Fred Sanders, Robert Sheets, Joanne Simpson, Harold Vanesee, John Weaver, Gilbert White, and Ray Zehr. We would also like to acknowledge the assistance of many individuals unknown by name to us at the other end of the phone line in various federal, state, and local agencies, particularly in south Florida. We would like to especially thank Rad Byerly, Kerry Emanuel, Chris Landsea, and Bob Simpson for assistance above and beyond the call of duty in reading and commenting on earlier versions of the manuscript. Chris Landsea also provided the spectacular color photos of Andrew's destruction that appear in the book and on the cover. Mary Downton, Justin Kitsutaka, Joe Eastman, and Judy Sorbie-Dunn expertly assisted with the figures. Jan Hopper, D. Jan Stewart, and Maria Krenz assisted at NCAR and the text was expertly produced at CSU by Dallas McDonald and Tara Pielke. Abi Hudlass, Clare Christian, and Mandy Collison at John Wiley steered the book to publication and we thank them for their help, guidance, and patience. We apologize to those people that we have inadvertently, but inevitably, forgotten to thank.

Each of these people has helped to make this book better than it otherwise would have been and we are very grateful for their help. Of course, all errors that remain in the text are the full responsibility of the authors.

ROGER A. PIELKE, Jr
Boulder, CO

ROGER A. PIELKE, Sr
Fort Collins, CO

18 February 1997

CHAPTER 1

Introduction: Science, Policy, and Hurricanes

1.1 THE HURRICANE: "A MELANCHOLY EVENT"

On 10 August 1856 residents of New Orleans, Louisiana experienced a moderate tropical storm accompanied by heavy rain and gusting winds. Damage was minimal and people paid little attention to the event. In the days following the storm, New Orleans residents learned that they had narrowly missed the impacts of a severe hurricane. The residents of Last Island, Louisiana, a barrier island community 25 miles southwest of New Orleans, were not as fortunate. A letter from a survivor of the Last Island Hurricane to the *Daily Picayune*, a New Orleans newspaper, relates the horror of the disaster (as quoted in Ludlam 1963, pp. 166–167).

> As one of the sufferers it becomes my duty to chronicle one of the most melancholy events which has ever occurred. On Saturday night [August 9] a heavy northeast wind prevailed, which excited the fears of a storm in the minds of many; the wind increased gradually until about ten o'clock Sunday morning, when there existed no longer any doubt that we were threatened with imminent danger. From that time the wind blew a perfect hurricane; every house on the island giving way, one after another, until nothing remained.

The survivor recounts how the extreme meteorological event quickly turned to human tragedy.

> At this moment every one sought the most elevated point on the island [about 6 feet above sea level], exerting themselves at the same time to avoid fragments of buildings, which were scattered in every direction by the wind. Many persons were wounded; some mortally. The water at this time (about 2 o'clock PM) commenced rising so rapidly from the bay side, that there could no longer be any doubt that the island would be submerged. The scene at this moment forbids description. Men, women, and children were seen running in every direction, in search of some means of salvation. The violence of the wind, together with the rain, which fell like hail, and the sand blinded their eyes, prevented many from reaching the objects they had aimed at.

The survivor next relates how the hurricane storm surge – a dome of water that accompanies the low pressure and high winds near the center of a hurricane – completely engulfed the low-lying barrier island, taking into the sea all of the island's structures and residents.

Many were drowned from being stunned by scattered fragments of the buildings, which had been blown asunder by the storm; many others were crushed by floating timbers and logs, which were removed from the beach, and met them on their journey. To attempt a description of this event would be useless.

The human suffering related to the storm did not end with the hurricane's departure, as opportunists pillaged the bodies of the hundreds of dead washed up on the beaches. In subsequent years the horror and lasting effects of the Last Island, or Isle Derniere, storm became part of the lore of Louisiana bayou country.

1.2 "WE NEED HELP": HURRICANE ANDREW IN SOUTH FLORIDA, AUGUST 1992

...FOR INTERGOVERNMENTAL USE ONLY...
TROPICAL DEPRESSION THREE DISCUSSION NUMBER 1 NATIONAL WEATHER SERVICE MIAMI FL 11 PM AST SUN AUG 16 1992
SATELLITE ANALYSTS AT BOTH SAB AND NHC HAVE BEEN CLASSIFYING THE TROPICAL WAVE OVER THE MID ATLANTIC FOR THE PAST COUPLE OF DAYS. THE DVORAK CI NUMBER HAS BEEN 2.0 FOR THE PAST 12 HOURS. METEOSAT IMAGES CONTINUE TO SHOW A BANDING TYPE PATTERN AND DEEP CONVECTION HAS INCREASED CLOSE TO THE CLOUD SYSTEM CENTER ... ENOUGH TO WARRANT ISSUING DEPRESSION ADVISORIES.
NMC AVIATION MODEL INDICATES THAT THE SYSTEM SHOULD REMAIN EMBEDDED WITHIN THE DEEP EASTERLIES ... AND ALL TRACK MODELS ARE IN GENERAL AGREEMENT. INITIAL MOTION IS 280/18 AND THIS SAME GENERAL MOTION SHOULD CONTINUE THROUGHOUT THIS FORECAST PERIOD.
THE AVIATION MODEL ALSO INDICATES STRONG 200 MB WINDS FROM THE EAST OVER THE SYSTEM ... WHICH SHOULD ALLOW ONLY SLOW STRENGTHENING. IF THE SHEAR IS LIGHTER THAN FORECAST ... MORE STRENGTHENING COULD OCCUR.

Tropical Cyclone Discussions such as this are prepared by the US National Hurricane Center (NHC) every six hours in order to provide decision-makers with up-to-date information on a storm's development and its track forecast (see Section 6.3.1 for further discussion). With this announcement, at 11:00 pm Sunday, 16 August 1992, the US National Hurricane Center alerted decision-makers in the US government that a typical tropical depression had formed from one of the about 60 tropical waves that originate off the West African coast (Avila and Pasch 1995). (See Section 1.3.1 for discussion of the categorization of tropical cyclones by intensity.) At this time there was little need for any public or private action, and perhaps not even awareness, as most tropical depressions fail to develop into hurricanes, and, of the hurricanes that do form, few threaten the United States. Yet "tropical depression three" was one which did fully develop into a hurricane and then strike the US coast. Less than eight days after the first discussion of "tropical depression three", south Florida was in the final hours of the landfall of Hurricane Andrew, the most powerful storm to strike the US since Camille in 1969. Tropical depression three survived the difficult process of hurricane growth to evolve from a loosely organized tropical system to become the most devastating storm in US history. The following sections tell the story of Hurricane Andrew in south Florida: not dealing comprehensively with the event, but rather giving the reader a sense of context for the event and a sampling of what happened in the communities of south Florida as Andrew approached, made landfall, and left.[1]

1.2.1 Forecast

On Monday, 17 August 1992, tropical depression number three was upgraded to tropical storm Andrew, the first named storm of the year. At this time, forecasters anticipated *"only gradual strengthening"* (Discussion Number 3). During the day on Tuesday, satellite imagery suggested that Andrew was weakening due to upper-level winds that were shearing the storm. On Wednesday, 19 August, the first reconnaissance flights flew into the storm to gather meteorological data. By late Wednesday afternoon, forecasters cautioned decision-makers that *"it is much too early to speculate whether Andrew will make landfall"* (Discussion Number 12). Forecasters concluded that the *"system still has to go through one more night of somewhat hostile winds aloft. If Andrew is not too damaged after that . . . there is a potential for strengthening for the remainder of the forecast period"* (Discussion Number 12). Meanwhile, coastal residents went about their normal routines: to those outside the forecast community, Andrew was not yet a tangible threat.

On Thursday morning, 20 August, reconnaissance flights indicated that the storm had fared poorly during the night. Forecasters noted that *"the struggle may be ending soon,"* but warned that *"satellite pictures are still impressive for a system that has gone through such a struggle"* (Discussion Number 15). Later that afternoon it had become apparent that the storm had survived the period

---

[1] The following sections rely primarily on the extensive coverage of the storm provided by *The Miami Herald*. The newspaper dedicated a number of special editions to the storm and in the weeks and months following the event published thousands of articles on Andrew's impact and the public's response. A search of the *Miami Herald* for the period 16 August to 31 December 1992 conducted resulted in a tabulation of close to a thousand articles on the storm and its impacts. The *Herald's* excellent coverage is a valuable resource for additional research into the storm's societal impacts. Of course, any narrative of the event cannot hope to convey the magnitude and severity of the disaster felt by the residents of Dade County, south of Miami.

of greatest upper-level wind shearing and forecasters anticipated that *"the environment is expected to be favorable for strengthening in the next 24 to 48 hours"* (Discussion Number 16). At this point Florida's State Emergency Operations Center (SEOC) was in a readiness mode. By Friday morning, 21 August, Andrew showed additional signs of strengthening, and forecasters expected the storm to reach hurricane strength within 24 hours. Figure 1.1 shows Andrew's storm track as the storm approached and crossed the south Florida coast.

At this point, there was little cause for alarm as NHC officials advised State Emergency officials that the storm would not likely threaten Florida until early in the next week (Koutnik 1993). At 11:00 pm Friday night forecasters made first reference to the potential landfall of Andrew: *"It should be noted that if the storm were to move a little faster than forecast . . . it could affect the northern Bahamas in 48 hours and somewhere along the U.S. Coast in 72 hours were it to continue on its present course."* (Discussion Number 21). At 5:00 am Saturday morning, 22 August, Andrew was classified as a Category 1 hurricane with 75 mile per hour (mph) winds. At 2:00 pm that afternoon reconnaissance planes left from Miami and San Juan, Puerto Rico to gather data on the storm.

As the weekend began, public decision-makers began taking action as Hurricane Andrew approached the Florida coast. For instance, that Saturday afternoon access to the Florida Keys, a part of Monroe County, was limited and tourists were dissuaded from attempting to visit. Bill Wagner, Monroe County's Director of Emergency Management, advised that

> We want no visitors in the Keys right now. Those that are in the Keys should head back. Those that are planning on coming to the Keys should cancel their plans immediately. There is a good chance that the southbound traffic on U.S. 1 [the only access to the Keys] will be blockaded on Sunday. (Quoted in Hancock and Faiola 1992)

Later that Saturday afternoon, forecasters prepared Florida decision-makers for the issuance of hurricane watches and warnings, noting that if the storm's increased forward speed continues then *"a hurricane watch will likely be issued for portions of the Florida Coast late tonite [sic] or early Sunday morning"* (Discussion Number 23). "At 2:30 pm Dr. [Robert] Sheets, Director of NHC, calls the SEOC and requests a NAWAS (National Warning System) conference call with all counties" (Koutnik 1993). At 5:00 pm Saturday a hurricane watch was issued for the Florida coast. Forecasters cautioned that:

> *All interests should be aware that landfall predictions are still nearly 48 hours away and precise points and times of landfall have considerable uncertainty. That*

Figure 1.1   Hurricane Andrew's storm track through the Atlantic and the Gulf of Mexico, 22–26 August 1992

5

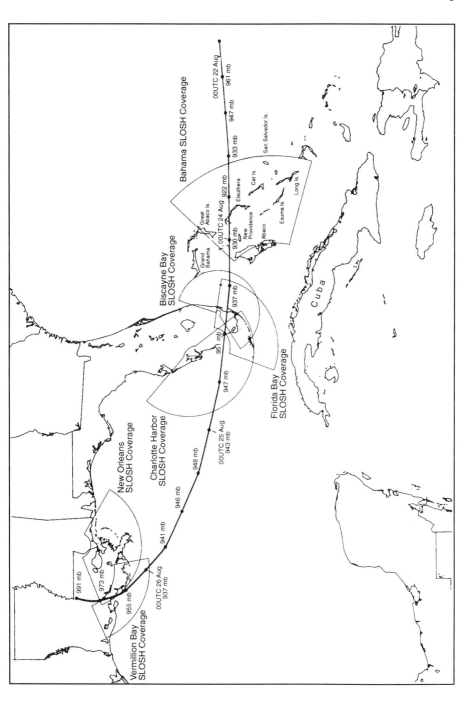

*is . . . a slight change of 3 to 5 degrees in direction can mean the difference
between the southern part of the watch area and the northern part of the watch
area . . . Do not focus on the exact track.* (Discussion Number 24)

Decision-makers along the entire Florida coast were advised to focus on
Andrew's steady approach.

At 11:00 pm Saturday night Andrew was a Category 3 storm 520 miles east
of Miami, and moving almost due west at 14 miles per hour. If Andrew
continued west at that rate, it would make landfall somewhere in south
Florida in a little over 37 hours, or at about noon on Monday. Dade County
officials scheduled a 6:00 am meeting for Sunday morning to discuss evacuation plans. Kate Hale, Dade County's Emergency Management Director,
advised that "Everyone from Washington on down is watching this very
closely. It's a matter of get ready, set, and in the morning, if we have to, GO!"
(Hancock and Faiola 1992). Officials in Broward County, immediately to the
north of Dade County, scheduled a similar meeting for 7:00 am. At the
meetings evacuation orders were issued based upon the NHC's issuance of
hurricane warnings for all of south Florida. Later Sunday afternoon hurricane
warnings were issued for portions of Florida's west coast as well as Lake
Okeechobee resulting in evacuation orders (Discussion Number 29).

In total, about 700 000 people of the one million who were ordered to
evacuate from Andrew's path did so. That left 300 000 people who were
ordered to evacuate, but did not.

People chose to stay in the evacuation zones for various reasons. For
instance, at 10:00 pm Sunday night, 23 August, 300 American and European
tourists were having a hurricane party in several beachfront Fort Lauderdale
hotels (Haner and Rafinski 1992). Fort Lauderdale City Manager George
Hanbury called the hotels and demanded to know why they had not yet
evacuated. With the majority of the evacuation completed by that time
through the use of private transportation contractors, city officials had to
recruit volunteers to drive school buses to evacuate the partiers. The next day,
Broward County Administrator Jack Osterholt blamed the tourists for poor
judgment: "There's no accounting for stupidity. That kind of behavior is
reprehensible. They obviously thought all of this was amusing – and they
risked the lives of some very brave people as a result" (quoted in Haner and
Rafinski 1992). A representative of the hotel chain observed that "It's hard to
force people out" who do not want to leave.

When asked before the storm why they refused to evacuate, many individuals expressed disdain for the storm, their machismo and stubbornness,
their ignorance of a hurricane's impacts, or concern about weathering the
storm in an unfamiliar location (Getter 1992a). One 81-year-old man refused
to leave his home near the beach: "This is the safest building on the Beach.
I've been here 24 years and been through 3 or 4 big ones. I'm staying right
here. I've got a wife that's strong and I'm strong" – an irony being that the

last "big one" to hit south Florida was Hurricane Betsy, 27 years earlier, and the last direct hit to metropolitan Miami occurred in 1950 (Landsea et al. 1996). Another person justified his decision not to evacuate in terms of his refusal to fear the storm: "I'm a native. I don't really worry about hurricanes. I'm going to stay indoors. It's just a panic from people who are new. I'm a Key Wester. I don't panic" (both quotes from Getter 1992a). In one mobile home park, in the hours before the storm, police asked those who refused to evacuate for the names of their next of kin "to be notified when we find your body in the rubble" after the storm (quoted in Van Natta 1992).

On Sunday, the SEOC was fully activated and FEMA activated its Regional Operations Center in Atlanta. FEMA also activated its Federal Response Plan and its Advance Emergency Response Team. At 8:00 pm Sunday night, the State of Florida submitted a request for a Presidential Disaster Declaration based on an expectation of disaster (Koutnik 1993).

1.2.2 Impact

At 4:35 am Monday, 23 August, Hurricane Andrew, the third most intense storm to strike the United States this century, made landfall over the Turkey Point area south of Miami. Figure 1.2 shows the path that Andrew took as it made landfall south of Miami (see also Figure 7.2).

Forecasters at the National Hurricane Center continued to issue forecasts as the storm battered their offices on the sixth floor of an eight-story office building across South Dixie Highway from the University of Miami. "*The center of Andrew is coming ashore in southern Dade County. Wind gusts to 138 mph have occurred at the National Hurricane Center*" (Discussion Number 32). As the storm's intensity increased, the NHC lost its radar. It was blown off the roof of the building and fell eight stories onto the property of a neighboring Holiday Inn. (Interestingly, the technology behind the NHC radar that was lost during Andrew was developed in the late 1950s, partly as a consequence of the severe hurricanes of the 1950s. With new "doppler" radar technologies in place, the old NHC radar was not replaced.)

As a result of Dade County's evacuation order, on Sunday evening 3500 people sought refuge at the North Miami Beach High School shelter, although the shelter was intended for only 2000 (Barry 1992). Almost half of the people at the shelter were elderly. Families, tourists, and people with nowhere else to go made up the rest. The crowded shelter tested people's interpersonal relationship skills. There were minor conflicts over cots and blankets reserved for the elderly as well as between nonsmokers and smokers. The refugees had exhausted the shelter's food supply during the evening meal, meaning that there would be no food available in the morning following the storm's passage. As the full fury of Andrew rapidly approached, several shelter officials made a daring run across town to replenish supplies. As the storm hit,

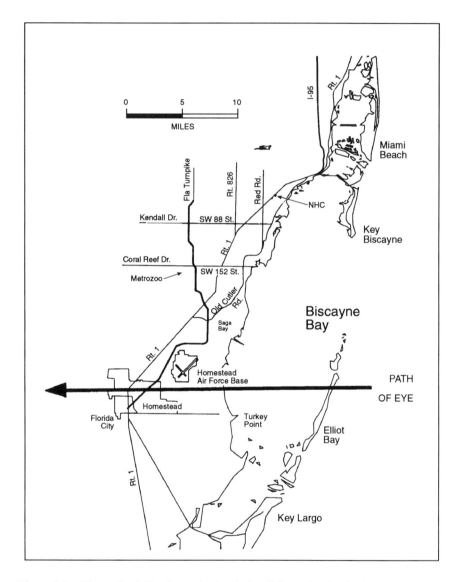

Figure 1.2   The path of Hurricane Andrew's landfall south of Miami, 24 August 1992

some refugees slept soundly, some worried whether they were going to make
it, and others tried to slip outside for a quick cigarette (Barry 1992).

   Although the North Miami Beach High School shelter was crowded beyond
capacity, many fewer people sought refuge in the shelters than had been
expected. According to one report, 43 000 people weathered the storm in
47 public shelters. Red Cross officials said that they would have been

overwhelmed had the shelters been refuge for the full 75 000 Dade County residents that were originally expected (USACE/FEMA 1993). In Broward County, about 26 000 people sought refuge in the public shelters, out of the 70 000 that were expected (Markowitz 1992).

One hero of the event was a television weatherman (Bodeker 1992). Brian Norcross of Miami's Channel 4 was on the air for about 22 hours straight, with only a few breaks. The safety of many people is attributed to the directions that he broadcast from a storage closet at Channel 4 (George 1992). According to Leonard Pitts, Jr, a staff writer for *The Miami Herald*:

> TV weatherman Bryan Norcross, broadcasting on radio station Y-100 probably saved our lives . . . "If you feel threatened," [Norcross] said, "don't be afraid to hunker down in an interior closet and shield yourself with mattresses." It sounded like such a foolish thing to do. The house shuddered. Water came through the front door and hit the back of my neck. "Honey," I said, "this may sound like a foolish thing to do, but . . ." Moments later, we had gathered food and water and heavy pillows from the couch, emptied a bedroom closet of clothes and shoes and taken refuge inside. We spent the remainder of the night there, sometimes listening by flashlight to updates from Norcross and his colleagues. (Pitts 1992)

In addition to where to seek refuge before the storm, Norcross informed people as to what supplies to stock, what to do with cars and pets, and the best routes to evacuate. After the storm Norcross was widely praised in south Florida and beyond.

Not all were fortunate enough to find safe haven from the storm. Lorenzo and Josefa Sopedra, both Cuban immigrants, took refuge in their poorly built farmhouse in south Dade County with their five children and five other people. At about 4:00 am their roof blew off, and all 12 of them rushed through the rain of debris and wind-driven missiles to a small storage trailer near the house. Within minutes the wind had begun to roll the trailer across a field. One after another each of them fell from the trailer as it split open. It finally disintegrated completely, exposing them to the raw power of the storm. Vidal Perez, who lived and worked on the farm, died of the wounds he sustained when he fell from the trailer; Sopedra's brother died as well (Garcia, Neal and Tanfani 1992). The two victims from the Sopedra farmhouse were among 14 people who lost their lives during the hours of Andrew's landfall in south Florida (DOC 1993).

As Andrew's winds diminished, residents of south Florida emerged from their homes and shelters to scenes of devastation. However, for forecasters at the NHC, Andrew's impacts were still of primary concern as they turned their attention to where Andrew would go next. At 9:00 am a hurricane watch was posted from Mobile, Alabama, to Sabine Pass, Texas, as the Gulf Coast braced for Andrew's second landfall.

The full fury of Andrew's impact in south Florida lasted less than five hours. In this short period lives were changed forever. A police officer from

Homestead, Ken Moore, was on duty during those five hours. He said that he would not do so again: "Next time, I'm evacuating" (*Miami Herald* 1992b). Another Homestead resident also refused to stay through the next one: "It was terrible. I've never been through a hurricane before and I'll never go through one again. I'm out of here. I'll board up and I'm gone" (quoted in Alvarez 1992a). Although Andrew will not be remembered for the number of lives that it took, it will be long remembered as a very close call by those hundreds of thousands of people who lived through it.

1.2.3 Response

Immediately following Andrew's rapid pass over south Florida, information of damage and casualties was difficult to come by. Some in the national media commented that at first the impacts did not appear as bad as expected, as first reports indicated that Miami Beach and downtown Miami were relatively intact. "We may have lucked out," said Miami Police Chief Philip Huber following a tour of Miami Beach immediately after the storm's exodus (Dewar 1992). Others were less sanguine. An anchor of the NBC Today Show, who was a former Miami resident and thus familiar with the area, cautioned the nation that there had been no reports from the communities south of Miami. Florida Governor Lawton Chiles announced that "We keep getting reports that Homestead is hit very, very hard. A certain percentage of it is no more. We think we're hit awfully hard in South Dade" (quoted in Markowitz 1992).

As information began to filter to the outside world from the center of Andrew's destructive path, the full magnitude of the storm's impact became apparent. In the first hours of the storm's aftermath, a National Guardsman described south Dade County in apocalyptic terms: "This place looks like ground zero after a nuclear blast – minus the radiation" (Elgiston 1992). A resident of south Dade County observed that "All it needs are fires and this would be a war zone" (Garcia, Neal and Tanfani 1992). Army Captain Johnny Dunlap agreed: "It looks like a war zone. It looks like this whole place has been under fire" (Slevin and Greene 1992). Andrew devastated the south Florida communities of Homestead, Leisure City, Goulds, Princeton, Naranja, and Florida City. For some, the reality of the destruction was difficult to comprehend. A resident of West Kendall, Miami noted that the scene through his shattered living room window "looks like what you see on CNN during a hurricane" (Dewar 1992). Nine hours after landfall, at 2:00 pm Monday, August 24, President George Bush declared a major disaster in Florida (FEMA 1993).

In the days following the storm, Dade County residents flooded neighboring counties looking for food, water, ice, disposable diapers, film (for insurance purposes), and cash (Penn and Evans 1992). According to a representative of Walgreen's drugstore, mobilization of supplies for after the event was extremely difficult compared to previous experience with Hurricane Hugo in South

Carolina in 1989, because two days after its Florida landfall Andrew was still a dangerous storm in the Gulf of Mexico. It was feared that the storm could perhaps bring destruction to New Orleans, Louisiana, or the Texas coast, making the distribution of supplies a difficult and uncertain challenge.

> We were prepared before this hit. The only problem this time is, with New Orleans and Texas threatened, three of our largest markets are hurt or about to be. Deciding where you need to ship, what you need to ship and getting it there is difficult. (Quoted in Penn and Evans 1992)

By Wednesday, August 26, Florida Power and Light (FPL) had restored power to about 43% of the 1 379 500 homes and businesses that had lost power in the storm (Dubocq 1992). Yet for many of those left without power, the restoration of service was as far off as five or six weeks, according to the best estimates of FPL officials. Power was the least of some people's problems. The pages of *The Miami Herald* served as a message board for friends and relatives trying to find one another. Along with names and phone numbers, the following message was typical: "Anyone who has seen my son or his girlfriend, please call me collect" (*Miami Herald* Staff 1992a). Others simply asked for help: "Need shoes and transportation, and soon a place to stay. Need carpentry job, my tools survived." Meanwhile, as residents of south Florida were about to begin a third day of coping with Andrew's impact, the storm was making landfall in rural Louisiana. Early Wednesday morning Andrew struck the Louisiana coast as a Category 3 storm.

The next day, Thursday, August 27, Kate Hale, Director of Dade County Emergency Operations, criticized the Federal response to the disaster and angrily asked: "Where the hell is the cavalry on this one? We need food, we need water, we need people" (Lyons and Merzer 1992). On Friday *The Miami Herald* announced to the world through an enormous front-page headline: "WE NEED HELP" (Figure 1.3). The Federal government responded quickly to the cry for help. President George Bush, in the final months of the presidential race, was sensitive to the feelings of residents of Florida, and the large number of electoral votes that the state carried. Tent cities for thousands, battlefield kitchens for 72 000 people daily, 600 000 meals-ready-to-eat from the Persian Gulf War, a field hospital, water, and blankets were the first resources to arrive. Hale was pleased at the response: "We are now seeing the federal response that we called for. We are starting to get the resources and personnel in" (Lyons and Merzer 1992). Federal government officials promised that military planes would arrive with supplies every 15 minutes for 36 hours.

The magnitude of the disaster, however, belied easy recovery. A week after the storm, Dade County was still in the throes of disaster, and the magnitude of response efforts created new problems. Roads, telephone lines, electricity and other infrastructure were in many cases jammed and inoperable. In Dade County, 70% of the 2400 stoplights were not working, compounding the

Figure 1.3   The front page of the *Miami Herald*, Friday, 28 August 1992. Reprinted with permission of *The Miami Herald*

traffic snarls caused by residents and sightseers (Slevin and Greene 1992). The confusion was felt by the various groups trying to facilitate recovery. In one instance, miscommunication between the National Guard and a volunteer project resulted in state workers being restricted from gathering in state parks to organize themselves and their volunteers. According to National Guard Major Gregory Moore, the confusion stemmed from the difficulties of such a large-scale response: "The confusion you see is not because people are going out screwing up. It's because we are overwhelmed with the amount of missions that we've been given" (Slevin and Greene 1992).

Meanwhile the storm's impact created demand for reconstruction and repair. "People are coming from all over the place," said a construction equipment operator who came from Tennessee looking for work following the storm. "There's a lot of money to be made here. I'll stay here even if I have to buy a tent to live in. They need my help here" (Alvarez 1992b). Another job-hunter from Georgia brought his family and said that "We'll stay a couple of years. There will be a lot of work here for a long time" (Alvarez 1992b). Opportunists sought to capitalize quickly on the demand for supplies essential to people's well-being. In the days following the storm, ice sold for $5 per bag, a can of tuna was $8, and bottles of water sold for $15 (Trausch 1992). Many individuals made a quick buck on the misfortune of others.

Several weeks into the recovery, people began asking about the next hurricane – a marked change from the years of complacency (Donnelly 1992). Andrew started with an "A", meaning that it was the first tropical storm of the year. It was conceivable that another hurricane or tropical storm could threaten the vulnerable south Florida area sometime later in the 1992 hurricane season. Kate Hale, Dade County's Emergency Director, expressed concern: "We have a plan for [another hurricane], but that plan is not going to work too well right now. Things aren't the way they used to be. We need to get ready. I'm very worried about it. I'm worried about a tropical storm coming" (quoted in Donnelly 1992). Of course, the odds of a second storm striking the south Florida coast were the same as they were before Andrew – it was public anticipation and expectations that had changed.

The influx of billions of dollars in federal aid in the days and weeks after the storm created a special set of problems. Government officials were criti-cized for allowing a private contractor to give out large bonuses – at the government's expense – for exceptional service during the recovery. Reports of kickbacks, bribery, and unauthorized overtime pay were commonplace. FEMA officials noted that such mishandling of the public's money was a common occurrence in the aftermath of disasters and was the consequence of both innocent mistakes and crass exploitation (Higham 1992a). To help facilitate the distribution of the billions of dollars in federal aid, the Florida State Community Affairs Office established a full-time "disaster office" at Miami International Airport. Private consultants were also hired to help. Yet, for the most part, difficulties in coordinating the large-scale recovery effort meant that sincere mistakes were difficult to avoid and opportunists had ample chances to exploit the disaster-aid system. As one participant noted: "We're all new at this, we're all learning and we're all working very hard" (Higham 1992a).

As time passed, Dade County slowly recovered and moved towards restoration. In mid-October 1992, the last of the 23 000 members of the US military left the south Florida area, ending the largest relief effort ever seen in the United States (Higham 1992b). Less than a week later, the last of the four tent cities that had housed Andrew's homeless closed down. It was two months after the storm's landfall (Hartman 1992). Residents of the tent cities moved to public shelters, rebuilt homes, or simply moved away. In November 1992, the night-time curfew which had restricted South Dade County residents to their homes, was lifted as a consequence of a lawsuit filed by a number of frustrated residents (Lyons 1992). South Dade County had spent almost three months in a state of pseudo-martial law.

1.3 DEFINING THE PROBLEM

An analysis of societal vulnerability to hurricanes in the United States helps to illuminate at least three interrelated problems. First, hurricanes threaten people and property along the US Atlantic and Gulf coasts. A better understanding of hurricanes and their societal impacts has the potential to contribute to efforts to reduce society's vulnerability to hurricane-related impacts. Second, hurricanes are a subset of a broader class of extreme weather events that threaten society. Lessons drawn from successes (and failures) in reducing vulnerability to hurricanes have the potential to contribute to efforts to reduce societal vulnerability to extreme weather events more generally. And third, understanding how society responds to the hurricane threat has the potential to contribute to the ongoing debate over US science policy regarding the efficacy of research supported with federal funds.

The differences between the impacts of the Last Island storm, almost 150 years ago, and Hurricane Andrew illustrate how much society has changed. Coastal population and property have expanded tremendously and so have efforts to prepare better for the extreme impacts of a hurricane. Yet some aspects of society's vulnerability to hurricanes remain much the same, if not worse. In 1980, Dr Neil Frank, Director of the National Hurricane Center, observed that since the turn of the century:

> . . . the United States has put men on the moon, orbited satellites to forecast the weather, and invented that miracle of modern civilization, the pop-top can. One might assume that our technological ingenuity has reduced or eliminated the risk of losing substantial numbers of lives in a hurricane. That is, however, not true. In fact, the hurricane peril has significantly increased. (Quoted in Baker 1980)

In 1986, the American Meteorological Society (AMS) issued a "Statement of Concern" entitled "Is the United States Headed for Hurricane Disaster?" Their answer to that question was an emphatic "Yes".

> We are more vulnerable to hurricanes in the United States now than we have ever been in our history. . . . This statement is a plea for the protection of the lives and property of United States citizens. If we do not move forward quickly in seeking solutions to the hurricane problem, we will pay a severe price. The price may be thousands of lives. (AMS 1986)

Recent experiences with Hurricanes Hugo (1989) in South Carolina, Andrew (1992) in Florida and Louisiana, Opal (1995) in Florida and Alabama, and

Fran (1996) in North Carolina seem to support the AMS warning. Although casualties associated with these storms were relatively low, they provide visceral evidence that the hurricane remains a "melancholy event".

1.3.1 The hurricane as an extreme meteorological event

The hurricane, one of the most powerful natural phenomena on the face of the Earth, is a member of a broader class of phenomena called cyclones (see Anthes (1982), Dunn and Miller (1964), Elsberry et al. (1987), and Simpson and Riehl (1981)). The term "cyclone" refers to any weather system that circulates in a counterclockwise direction in the Northern Hemisphere and in a clockwise direction in the Southern Hemisphere. "Tropical cyclones" form over ocean waters of the tropics (the area on the Earth's surface between the Tropic of Capricorn and the Tropic of Cancer, 23 degrees 27 minutes south and north of the Equator, respectively) and subtropical waters, on occasion. Extratropical cyclones, for comparison, form as a result of the temperature contrast between the colder air at higher latitudes and warmer air closer to the Equator. Extratropical storms form over both the ocean and land.

The meteorological community uses a number of terms to classify the various stages in the life cycle of tropical cyclones (adapted from Neumann 1993; Neumann, Jarvinen and Pike 1987). In this book, we adopt the definitions of tropical cyclones as used in the Atlantic Ocean basin.

Tropical low   A surface low-pressure system in the tropical latitudes.

Tropical disturbance A tropical low and an associated cluster of thunderstorms which has, at most, only a weak surface wind circulation.

Tropical depression A tropical low with a wind circulation of sustained 1-minute surface winds of less than 34 knots (kt) (39 miles per hour (mph), 18 meters per second (m/s) circulating around the center of the low). (Most countries use a 10-minute average to define sustained winds. A knot (i.e. a nautical mile per hour) equals about 1.15 miles per hour. A nautical mile is the length of one minute of arc of latitude.)

Tropical storm   A tropical cyclone with maximum sustained surface winds of 34 to less than 64 kt (39 to 74 mph, 18 to 33 m/s).

Hurricane   A tropical cyclone with sustained surface winds 64 kt (74 mph, 33 m/s) or greater. (In the Pacific Ocean west of the International Date Line, hurricanes are called typhoons. They are the same phenomenon.)

Hurricanes are classified by their damage potential according to a scale developed in the 1970s by Robert Simpson, a meteorologist and director of the National Hurricane Center, and Herbert Saffir, a consulting engineer in Dade County Florida (Simpson and Riehl 1981). The Saffir/Simpson scale was developed by the National Weather Service to give public officials information on the magnitude of a storm in progress and is now in wide use by producers and users of hurricane forecasts. The scale has five categories, with Category 1 the least intense hurricane and Category 5 the most intense. Table 1.1 shows the Saffir/Simpson scale and the corresponding criteria for classification.

1.3.2 Hurricanes in North American history

The word "hurricane" derives from the Spanish "huracán", itself derived from the dialects of indigenous peoples of the Caribbean and Latin America (Dunn and Miller 1964). "Hunraken" was the name of the Mayan storm god, and "Huraken" was the god of thunder and lightning for the Quiche of southern Guatemala (Henry et al. 1994). The Tainos and Caribe tribes of the Caribbean called their God of Evil by the name "Huracan". Other indigenous dialects included words such as "aracan", "urican", and "hurivanvucan" to refer to "Big Wind". The deification of the hurricane and the connection of indigenous words for the phenomena with evil and violence is an indication that hurricanes had a significant impact on the lives of many indigenous peoples of the Caribbean and Latin America.

The historical record of documented hurricane events begins with the European conquest of North America. Columbus, in his four voyages to North America, experienced direct contact with an Atlantic hurricane only in his fourth voyage. Meteorological historian David Ludlam notes that Columbus' good fortune in his first voyage leads one to wonder "what the course of history in the West Indies might have been if, in the autumn of 1492, a full-blown tropical storm had dashed the frail craft of the Admiral's fleet to the bottom of the sea or flung them shipwreck on some tiny cay" (Ludlam 1963). Others did not experience such good fortune. In 1667, a "dreadful hurry cane" struck North Carolina and Virginia, causing extensive damages (Ludlam 1963). Shakespeare's play, The Tempest, was loosely based on reports of a 1609 hurricane near Bermuda that sunk the vessel Sea Venture and stranded the passengers, including John Rolfe, future husband of Pocahontas, on the island for 10 months. This storm's movement was among the first successfully anticipated by the colonists. During the course of the storm's trek through the Caribbean, a skipper in the Royal Navy cautioned the British fleet to move out of the storm's path, based on his experience with the movement of past hurricanes. During the 1700s and 1800s numerous coastal locations were struck by severe hurricanes. Charleston, South Carolina, New Orleans, Louisiana, and Boston, Massachusetts, were particularly hard hit a number of times. In 1772 in the West Indies, a teenaged Alexander Hamilton wrote about a hurricane's impact for a local newspaper. His writing caught the attention of the local gentry, who then raised money to

Table 1.1 Saffir/Simpson Hurricane Scale (after Hebert, Jarrell and Mayfield 1993; Saffir 1995, personal communication)

Category	Central pressure (millibars)	(inches)	Winds (mph)	$(m\ s^{-1})$	Surge (feet)	(meters)*	Damage	North American examples
1	≥980	≥28.92	74–95	33–42	4–5	1–1.5	minimal	Agnes 1972
2	965–979	28.50–28.91	96–110	43–49	6–8	2–2.5	moderate	Cleo 1964
3	945–964	27.91–28.49	111–130	50–58	9–12	3–3.5	extensive	Fran 1996
4	920–944	27.17–27.90	131–155	59–69	13–18	4–5.5	extreme	Andrew 1992
5	<920	<27.17	>155	>69	>18	>5.5	catastrophic	Camille 1969

* Rounded to nearest 0.5 m.

Table 1.2   The repeating six-year list of names of Atlantic tropical cyclones, given when a cyclone reaches tropical storm strength. The letters Q, U, X, Y, and Z are not used. 2003 will use the names of 1997 (except those that are retired), 2004 will use those from 1998, and so on

1997	1998	1999	2000	2001	2002
Ana	Alex	Arlene	Alberto	Allison	Arthur
Bill	Bonnie	Bret	Beryl	Barry	Bertha
Claudette	Charley	Cindy	Chris	Chantal	Cesar
Danny	Danielle	Dennis	Debby	Dean	Dolly
Erika	Earl	Emily	Ernesto	Erin	Edouard
Fabian	Frances	Floyd	Florence	Felix	Fran
Grace	Georges	Gert	Gordon	Gabrielle	Gustav
Henri	Hermine	Harvey	Helene	Humberto	Hortense
Isabel	Ivan	Irene	Isaac	Iris	Isidore
Juan	Jeanne	Jose	Joyce	Jerry	Josephine
Kate	Karl	Katrina	Keith	Karen	Kyle
Larry	Lisa	Lenny	Leslie	Lorenzo	Lili
Mindy	Mitch	Maria	Michael	Michelle	Marco
Nicholas	Nicole	Nate	Nadine	Noel	Nana
Odette	Otto	Ophelia	Oscar	Olga	Omar
Peter	Paula	Philippe	Patty	Pablo	Paloma
Rose	Richard	Rita	Rafael	Rebekah	Rene
Sam	Shary	Stan	Sandy	Sebastien	Sally
Teresa	Tomas	Tammy	Tony	Tanya	Teddy
Victor	Virginie	Vince	Valerie	Van	Vicky
Wanda	Walter	Wilma	William	Wendy	Wilfred

send him to the mainland colonies to further his education, thus setting the stage for his political career.

Tropical storms were once named after the particular "saint's day" that fell nearest the hurricane event (Tannehill 1952). For instance, "Hurricane Santa Ana" hit Puerto Rico on 26 July 1825 (see Rodriguez 1995). Today, tropical cyclones are "named" when they reach tropical storm strength. According to one explanation, this practice dates to the 1950s, following the publication of George R. Stewart's *Storm*, a book that featured a forecaster who named storms (Williams 1992). Another explanation is that the hurricane-naming convention began with a military radio operator who during World War II ended each hurricane warning singing "Every little breeze seems to whisper Louise," prompting the naming of a particular hurricane Louise (Henry, Portier and Coyne 1994). Whatever the origin, the practice caught on because it proved useful in identifying different storms that existed simultaneously. The personification of the extreme event was also found to be a valuable practice by the various user communities. Until 1979, tropical storms were given only women's names in English. In 1979 forecasters began to use men's, French, and Spanish names as well. Table 1.2 shows the repeating six-year list of names assigned to tropical cyclones in the Atlantic put together by the World Meteorological Organization and also found at the National Hurricane Center's website at www.nhc.noaa.gov/names.html (see Appendix D for a list of names used around the world). Hurricanes which cause significant damage or are particularly memorable, such as Andrew (1992), Camille (1969), or Gilbert (1988), are retired and those names are not used again. Table 1.3 lists the retired hurricanes up to 1995 and notes the death and damages associated with each. The deadliest and costliest hurricanes to impact the United States are listed in Appendix B.

1.3.3 Ten notable hurricanes of the past century

Each of the hurricanes that have had their names retired since 1954 are notable for the extreme impacts on the affected communities or regions. Indeed, many hurricanes that occurred prior to 1950 are also notable for their societal impacts. Rappaport and Fernandez-Partagas (1995), Ludlam (1963), and Tannehill (1952) describe significant historical storms of the period 1900–1950. However, a number of the hurricanes of the past century deserve special mention due to the significance of their impacts on the United States.

Galveston 1900

On 8 September 1900, tremendous waves began breaking on the beaches of Galveston, Texas, drawing curious spectators to the beach from the town (Dunn and Miller 1964). As the barometer of Dr I.M. Cline, a meteorologist, dropped rapidly, eventually reading 931 mb, he began to warn many residents of an approaching storm. Though many heeded his warning, others went to the beach to witness the high surf. It is commonly believed that 6000 people died, but recent research suggests that as many as 12 000 people, including Dr Cline's wife, lost their lives due to the high winds and 20 foot storm surge that devastated the city (Rappaport and Fernandez-Partagas 1995). The storm was the most deadly natural disaster in United States history (and remained so up to at least 1996).

Miami 1926

A powerful hurricane made landfall over Miami Beach at midnight, 18 September 1926 (Rosenfeld 1996). A Weather Bureau meteorologist recorded a low pressure reading of 27.61 inches of mercury, a Category 4 storm. The storm surge was about 8 feet at the Miami waterfront, and about 9 feet on Miami Beach. The largely unexpected storm left over 200 fatalities in its wake and hundreds of millions of dollars of damage (Dunn and Miller 1964). In another south Florida storm two years later, more than 1800 people drowned when Lake Okeechobee overflowed due to strong hurricane winds. A result of the 1928

Table 1.3   Thirty-nine "retired" Atlantic hurricane names through 1995. Damage cost figures are in 1990 dollars. Sources: NOAA Fact Sheet; Hebert, Jarrell and Mayfield (1993); Rappaport and Fernandez-Partagas (1995); C. Landsea, personal communication, 1997; R.H. Simpson, personal communication

Year	Name	Location	Notes (US costs (1990 $) and total casualties, etc.)
1954	Carol	Coastal areas north of the Carolinas to Long Island, inland along its track north into Canada, mid-Atlantic states	$2.37 billion, 60 deaths
1954	Hazel	Antilles, North and South Carolina	$1.44 billion, 1000 deaths
1955	Connie	North Carolina	25 deaths
1955	Diane	Northeast US	$4.20 billion, 184 deaths
1955	Ione	North Carolina	$444 million
1955	Janet	Lesser Antilles, Belize, and Mexico	538 deaths
1957	Audrey	Louisiana and North Texas	$696 million, 550 deaths
1960	Donna	Bahamas, Florida, and Eastern US	$1.82 billion, 364 deaths
1961	Carla	Texas	$1.93 billion, 46 deaths
1963	Flora	Haiti and Cuba	8000 deaths
1964	Cleo	Lesser Antilles, Haiti, Cuba, Southeast Florida	$595 million, 213 deaths
1964	Dora	Northeast Florida	$1.16 billion
1964	Hilda	Louisiana	$579 million, 304 deaths
1965	Betsy	Bahamas, Southeast Florida, Southeast Louisiana	$6.46 billion, 75 deaths
1966	Inez	Lesser Antilles, Hispaniola, Cuba, Florida Keys, Mexico	1000 deaths
1967	Beulah	Antilles, Mexico, South Texas	$844 million; the largest number of tornadoes (115) ever associated with a hurricane
1969	Camille	Louisiana, Mississippi, and Alabama	$5.24 billion, 256 deaths
1970	Celia	South Texas	$1.56 billion
1972	Agnes	Florida, Northeast US	$5.24 billion, 122 deaths
1974	Carmen	Mexico, Central Louisiana	$380 million
1975	Eloise	Antilles, Northwest Florida, and Alabama	$1.08 billion
1977	Anita	Mexico	
1979	David	Lesser Antilles, Hispaniola, Florida, and Eastern US	$487 million, 2000 deaths
1979	Frederic	Alabama and Mississippi	$3.50 billion
1980	Allen	Antilles, Mexico, and South Texas	$411 million, 249 deaths
1983	Alicia	North Texas	$2.39 billion, 21 deaths
1985	Elena	Mississippi, Alabama, and Western Florida	$1.39 billion, 4 deaths

Table 1.3 *continued*

Year	Name	Location	Notes (US costs (1990 $) and total casualties, etc.)
1985	Gloria	North Carolina and Northeast US	$1.00 billion
1988	Gilbert	Lesser Antilles, Jamaica, Yucatan Peninsula, and Mexico	327 deaths; lowest central pressure (888 mb) ever recorded
1988	Joan	Curacao, Venezuela, Colombia, and Nicaragua	216 deaths; crossed into Pacific and was renamed Miriam
1989	Hugo	Antilles and South Carolina	$7.16 billion, 56 deaths
1990	Diana	Mexico	96 deaths
1990	Klaus	Martinique	
1991	Bob	North Carolina and Northeast US	$1.5 billion
1992	Andrew	Bahamas, South Florida, and Louisiana	>$25 billion
1995	Luis	Leeward Islands	$2.5 billion, 16 deaths
1995	Marilyn	Virgin Islands	$1.5 billion, 8 deaths
1995	Opal	Mexico, Florida	$3 billion, 59 deaths
1995	Roxanne	Mexico	$1.5 billion, 14 deaths

storm was the construction of a levee around Lake Okeechobee to protect residents from future events. The levee still stands today.

Chesapeake–Potomac 1933

On 23 August 1933 a Category 3 storm made landfall near Nags Head, on North Carolina's Outer Banks (Cobb 1991). The storm's center moved northwest on a track that took it west of Norfolk, Virginia, Washington, DC, Baltimore, Maryland, and Atlantic City, New Jersey. The 1933 storm is significant as it was the only hurricane in this century directly to strike the Chesapeake–Potomac area. The storm caused extensive property damages along the Mid-Atlantic coast, deep in the throes of the Depression, and 47 people lost their lives. The hurricane provided an unexpected benefit to residents of Ocean City, Maryland, by cutting an inlet between Sinepatuexent Bay and the Atlantic Ocean (Corddry 1991). Residents had been lobbying Congress for several years for funds to create the harbor. The hurricane created the channel in one day.

A consequence of the numerous hurricanes from the mid-1920s through the mid-1930s was Congressional and Presidential action to restructure the hurricane warning service (NOAA 1993). Such action in the aftermath of catastrophe has been typical of the process of development of hurricane policies throughout the past one hundred years.

New England 1938

A rapidly moving storm tore across New England on 21 September 1938 (Pierce 1939). The storm, called by some "The Long Island Express" due to its forward speed of 60–70 mph, made landfall over Long Island, Connecticut, Massachusetts, and Rhode Island (Coch and Wolff 1990; Pierce 1939). The storm was the most severe to strike the area in at least 100 years, although major hurricanes had made landfall in New England in 1635, 1788, 1815, and 1869 (Foster and Boose 1992; 1995). Debate exists over whether it was a Category 2 or 3 storm. One wind gust was reported at Blue Hill, Massachusetts (outside Boston) at 186 mph (K. Emanuel 1997, personal communication). More than 600 people were killed and thousands were injured. Extensive forest damage occurred as far north as northern New Hampshire and Vermont (Foster 1988). At the time the storm "held the all-time record for storm property damage in the United States, and probably the world as well" (Dunn and Miller 1964). The storm's size was also large, with its radius of sustained hurricane-force winds over the ocean at landfall as great as 106 miles on its eastern side (170 kilometers; Boose, Foster and Fluet 1994). Results from that paper also estimate that hurricane winds in gusts extended towards to the right of the track of the storm to 144 miles (230 kilometers) over both the land and water. A 1990 study argued that "a recurrence of a Category 2 storm such as this 1938 hurricane, on the now highly urbanized Long Island shoreline would lead to greater property damage than Hugo caused in South Carolina [>$7 billion]". Furthermore, a Category 3 or 4 hurricane "will result in far greater damage than has ever been experienced on Long Island" (Coch and Wolff 1990).

Hazel 1954

Hazel moved inland over North Carolina and South Carolina on 15 October 1954 as a Category 4 hurricane after taking an erratic path through the Caribbean and along the southeast US Atlantic coast. "Normally hurricanes dissipate or at least lose considerable intensity when they move inland. Hurricane Hazel did neither" (Dunn and Miller 1964). Hazel joined with another storm system to devastate inland communities from Virginia to Ontario, Canada. Washington, DC experienced its strongest winds ever recorded. In total, Hazel resulted in more than 90 deaths in the US and damage amounting to hundreds of millions of dollars.

    One interesting consequence of Hazel's impact was to provide impetus for increased federal support of hurricane research. According to Robert Simpson, director of the National Hurricane Research Laboratory from 1956 to 1958, Hazel and a number of other severe US East Coast hurricanes (Carol, Hazel, Edna, Diane, and Ione, which resulted in damages of close to $10 billion (1990 $) (Hebert, Jarrell and Mayfield 1993) in 1954 and 1955, motivated Congress to increase funding for hurricane research.

> The Weather Bureau's efforts to obtain budgetary support to expand investigations of [hurricanes] fell on deaf ears in Congress until the siege of [hurricanes] Carol, Edna, and Hazel in 1954 supplied the imperative for support of a more comprehensive research effort than had ever before been seriously considered. (Simpson 1980)

Weather Bureau funding for research quintupled in three years. Again, there is a connection between a catastrophe and a series of policy responses.

Camille 1969

On 17 August 1969 Hurricane Camille made landfall over Mississippi, Alabama, and southeastern Louisiana (Simpson et al. 1970). With wind gusts of up to 200 mph and a storm surge of 25 feet, Camille, a Category 5 storm, was the most powerful hurricane to directly strike the US coastline in the 20th century.[1] One woman, attending a "hurricane party" in a hotel near the coast, was swept more than 12 miles inland by the storm surge. She was the only survivor of 23 fellow party-goers at the hotel, which was completely destroyed (NOVA 1993). Camille continued north along the Appalachian Mountains and caused massive flooding in central Virginia. All told, Camille left in its wake 256 people dead and more than $5.6 billion in damages (1994 $).

Frederic 1979

Between Agnes (1972) and Hugo (1989), the most costly hurricane to strike the US coast was Frederic, a Category 3 storm that hit Alabama and Mississippi in 1979. The storm destroyed much of the island of Gulf Shores, Alabama, and resulted in over $3.9 billion (1994 $) in damages. In testimony before Congress in 1992, Robert Sheets, director of the National Hurricane Center, used the Frederic experience to illustrate the irrationality of coastal development (SCBHUA 1992).

> Prior to Hurricane Frederic, there was one condominium complex on Gulf Shores, Alabama. Most of the homes were single, individual homes built behind the sand dunes. . . . Today, where there used to be one condominium, there are now 104 complexes – not units, complexes – on Gulf Shores, Alabama. . . . Have we learned?

---

[1] The 1935 Florida Keys hurricane had a lower minimum central pressure than Camille (892 mb versus 909 mb). The two hurricanes are the only Category 5 storms to make landfall in the United States in the 20th century. Hurricane Gilbert (1988) holds the record for lowest minimum central pressure, 888 mb, over the waters of the Gulf of Mexico. On 11 October 1846 a hurricane crossed the Florida Keys with a minimum pressure estimated as 908 mb (Schwerdt, Ho and Watkins 1979).

Alicia 1983

Alicia in August 1983 was the first hurricane to make landfall in the US since 1980. Alicia made landfall as a Category 3 storm over Galveston Island, Texas. The storm carried with it a 15 foot storm surge and spawned several dozen tornadoes during landfall. Landfall at Galveston came as somewhat of a surprise as the storm was poorly forecasted. Residents were fortunate to have in place the sea wall built following the 1900 disaster.

> Hurricane Alicia in 1983 had the potential to produce a catastrophe. It formed as a weak storm in the Gulf of Mexico in an environment that did not appear favorable for strengthening. Local officials in Galveston decided against complete evacuation of coastal areas. When Alicia strengthened significantly in the 18 hours before landfall, it was too late to totally evacuate the threatened area. Large loss of life was averted by the presence of a 15-foot sea wall that was built to protect the city following the record disaster of 1900. . . . Few coastal locations have such massive seawalls. If a similar situation occurred in an unprotected area, resulting casualties could number in the hundreds or thousands. (AMS 1986)

Subsequently, Alicia moved inland, passing directly over the Houston, Texas, area causing extensive damage. The storm resulted in 21 deaths and $2.8 billion (1994 $) in hurricane-related damages, at the time making it the costliest disaster in Texas history (Eagleman 1990).

Hugo 1989

When Hurricane Hugo made landfall just north of Charleston, South Carolina, on 22 September 1989, it was the first Category 4 (or higher) storm to strike the US coast since Camille (1969) 20 years earlier. Prior to its landfall on the US Atlantic coast, the storm had directly hit the US Virgin Islands and Puerto Rico. The storm cost 49 people their lives, and resulted in more than $9 billion in total damages, with more than $7 billion due to damage on the US mainland (DOC 1990). Although Charleston, South Carolina, escaped the full fury of the storm, the city suffered extensive damage. Loss of life and property could have been much worse had Hugo made landfall just a little south of where it did.

Andrew 1992

After passing over the island of Eleuthra, in the Bahamas, Hurricane Andrew made landfall over Dade County, Florida, on 24 August 1992 and then moved into the Gulf of Mexico before making a second landfall three days later over rural Louisiana. The storm was one of the most costly natural disasters in US history, the only comparable disaster events in damages being the Midwest Floods of 1993 and the Northridge Earthquake in Los Angeles in January 1994. The Category 4 storm, with a minimum pressure of 922 mb, resulted in an estimated $25–40 billion in damages, most of which occurred in Dade County, south of Miami (Sheets 1994). As with Hurricane Hugo, the worst magnitude disaster did not occur. Studies conducted by the insurance industry and *The Miami Herald* suggest that had Andrew made landfall only 20 miles farther north, damages could have been more than twice as costly (DOC 1993; see also Figure 7.2). Hurricane Andrew precipitated a reconsideration of a number of hurricane-related policies, including those of the hurricane insurance industry as well as building code practices in south Florida and elsewhere.

1.3.4 Extreme weather events

The hurricane is part of a broader class of extreme weather phenomena that threaten the United States. Other phenomena include winter storms (e.g. snow, sleet, freezing rain, and freezes), thunderstorms (e.g. tornadoes, heavy rains, lightning, wind, and hail), extreme precipitation (e.g. flood and flash floods), and windstorms. According to one study, insurance claims due to extreme weather events for the period 1950 to 1989 account for $66.2 billion (1991 $) in insured losses during the 50-year period, with about half due to hurricanes (Changnon and Changnon 1992). (It is important to point out that insurance losses represent only a fraction of the total monetary and societal impact of extreme weather events.) For comparison, over the same period, one study (Landsea 1991) estimated *total* monetary losses due to hurricanes to be about twice the *insured* losses due to hurricanes. Figure 1.4 shows, for the period 1984–1993, insurance payouts for various disasters (BTFFDR 1995). Hurricanes are the most significant. Floods are not included in this pie chart because they are typically not insured by the private sector, but instead through the National Flood Insurance Program.

Tables 1.4, 1.5 and 1.6 show data kept by the National Weather Service on the economic damages and casualties associated with weather in the United States for 1992–1994. (The data presented here may not agree with other sources because of different methodologies used to compute damages and estimate deaths; see Chapter 5 for discussion.) In 1992 the total losses were more than $38 billion and in 1993, the losses were more than $28 billion, both of which were considerably more than 1994 with $4.4 billion (all current year dollars). However, 1994 saw more deaths and considerably more injuries than the previous two years. A growing body of evidence suggests that the US is more vulnerable to weather damages and casualties; consequently the losses of 1992 and 1993 and casualties of 1994 may be more typical in the near future (see, e.g., Pielke et al. 1997).

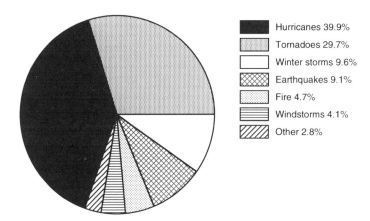

Figure 1.4   Insurance payouts for various disasters for 1984–1993. Source: BTFFDR (1995)

While the hurricane threat to the US East and Gulf coasts remains significant, past successes in reducing societal vulnerability to hurricanes (particularly the threat to human life) provide a significant base of experience that may serve as a case study of how scientific information can play a more useful role in preparation, mitigation, and response efforts to extreme weather events. In this regard, experience with hurricanes can provide important lessons for efforts to reduce society's vulnerability to weather. Looking to the future, communities that respond successfully to the hurricane threat can provide opportunities to demonstrate how current and future weather research might be leveraged to reduce societal vulnerability to a broader range of extreme weather events.

1.3.5 Science in service to society

In recent years, in the US as well as in other countries, the contributions of scientific research to the ameliorization of societal problems have faced close scrutiny. For instance, some members of Congress, both Republicans and Democrats, have called upon science to demonstrate the societal benefits that are often promised in efforts to secure federal funding (Byerly 1995). In light of changes in the environment of US science policy, it is likely that sustained federal support of scientific research, including weather research, will be a function of a particular program's performance measured against the claims made to Congress by its supporters (Pielke and Glantz 1995). It seems clear that US science policy is well into a period of change, with public accountability of science and research efficacy with respect to societal problems comprising the

Table 1.4   A summary of 1992 weather events, deaths, injuries, and damage costs (National Weather Service, unpublished data)

Weather event	Deaths	Injuries	Amount of damage ($ millions)
Convection			
Lightning	41	266	16.0
Tornado	39	1300	764.7
Thunderstorm winds	13	282	266.2
Hail	0	71	532.9
Extreme temperatures			
Cold	14	2	458.3
Heat	8	77	21.2
Flood			
Flash flood	55	41	428.3
River flood	7	207	262.6
Marine			
Coastal storm	0	5	31.1
Tsunami	0	0	0.0
Tropical cyclone			
TS/Hurricanes	27	298	33 611.3
Winter			
Snow/blizzard	43	125	18.1
Ice storm	16	252	9.9
Avalanche	5	2	0.0
Other			
Drought	0	0	1 780.4
Dust storm	3	10	0.2
Rain	2	0	0.5
Fog	14	173	1.6
High winds	15	44	44.4
Water spout	0	0	0.0
Fire weather	0	7	106.2
Mud slide	2	0	0.2
Other	4	77	0.2
Totals	308	3239	38 395.4

principal dimensions of the transformation (cf. Brunner and Ascher 1992; Byerly 1995; Byerly and Pielke 1995).

Many in the science community recognize that change is unavoidable. The National Academy of Sciences has proposed "a renewed and strengthened covenant between science, technology, and society" (NAS 1993). Neal Lane, Director of the National Science Foundation, testified before Congress in 1995 that one of the goals of "NSF in a Changing World" was the "discovery, integration, dissemination, and employment of new knowledge in service to society" (Lane 1995). However, change is not without enemies. For example, one scientist responded to "science bashers" on the "anti-science warpath", arguing that reconsideration of the "rules of research" runs the risk of

Table 1.5 A summary of 1993 weather events, deaths, injuries, and damage costs (National Weather Service, unpublished data)

Weather event	Deaths	Injuries	Amount of damage ($ millions)
Convection			
Lightning	43	286	32.5
Tornado	33	974	368.4
Thunderstorm winds	23	458	348.7
Hail	0	20	336.9
Extreme temperatures			
Cold	18	1	341.2
Heat	20	66	75.4
Flood			
Flash flood	51	48	408.7
River flood	52	45	20 880.7
Marine			
Coastal storm	0	0	0.7
Tsunami	0	0	0.0
Tropical cyclone			
TS/Hurricanes	2	1	15.0
Winter			
Snow/blizzard	58	501	584.2
Ice storm	8	92	18.4
Avalanche	1	0	0.0
Other			
Drought	1	0	514.0
Dust storm	0	1	0.0
Rain	0	14	3 299.8
Fog	5	9	0.2
High winds	40	129	231.1
Water spout	1	0	0.4
Fire weather	3	239	950.0
Mud slide	2	0	0.0
Other	11	1	25.0
Totals	372	2885	28 431.3

"ruining a priceless institution" (Kleppner 1993). Change has stimulated much debate in the US science policy community.

The hurricane threat to the US Gulf and Atlantic coasts provides the scientific community with an opportunity to demonstrate tangible societal benefits that are directly related to scientific research. Yet demonstration of benefits is often a difficult analytical challenge in practice, because "the path from scientific research to societal benefits is neither certain, nor straight" (Brown 1992). As one expert has noted, "adverse weather events by themselves can be devastating for society, but their effects are often exacerbated by economic, political, and social decisions made, in many instances, long before

Table 1.6   A summary of 1994 weather events, deaths, injuries, and damage costs (National Weather Service, unpublished data)

Weather event	Deaths	Injuries	Amount of damage ($ millions)
Convection			
Lightning	69	484	47.8
Tornado	69	1067	518.8
Thunderstorm winds	17	315	270.2
Hail	0	37	165.4
Extreme temperatures			
Cold	52	182	50.7
Heat	29	116	1.0
Flood			
Flash flood	59	33	533.7
River flood	32	14	386.7
Marine			
Coastal storm	1	0	2.9
Tsunami	0	0	0.0
Tropical cyclone			
TS/hurricanes	9	45	426.4
Winter			
Snow/blizzard	23	488	240.1
Ice storm	6	2189	903.1
Avalanche	2	13	0.0
Other			
Drought	0	0	225.0
Dust storm	1	19	0.0
Rain	4	1	281.8
Fog	3	99	0.2
High winds	12	61	42.0
Water spout	0	0	0.2
Fire weather	0	2	342.4
Mud slide	0	0	2.6
Other	0	0	0.0
Totals	388	5165	4441.0

those events take place" (Glantz 1978). Thus, while research holds much potential to contribute to reducing societal vulnerabilities to hurricanes, if that potential is to be realized in practice, care must be taken to understand such research in its broader political and social contexts.

In the broader context of US science policy, demonstrations and analyses of the use and value of hurricane research in reducing societal vulnerability have not found a broad audience (nor, for that matter, very many "players"). The observation of one hurricane expert in connection with aircraft reconnaissance and tropical cyclone forecasting has general relevance for understanding the lack of assessments of the use and benefits of hurricane research.

While relying on reconnaissance for more than 40 years, most American [tropical cyclone] forecasters and researchers have not felt the need to make quantitative studies of just how beneficial aircraft reconnaissance has been in order to justify its continuation. Research is now belatedly beginning to focus on this subject. (Gray, Neumann and Tsui 1991, p. 1981)

There are at least three reasons why the successes of hurricane research have not reached the broader science policy community. First, because for many years research funding flowed relatively freely, hurricane researchers have had little incentive to conduct assessments of the use and value of their research. Second, members of the hurricane research community are poorly placed to argue for the worth of their research, as some policymakers may view such arguments as self-serving, no matter how meritorious. Finally, the question of the use and value of hurricane research is a difficult analytical question that involves the assessment of the process of decision-making well beyond the contribution of scientific research, to include scientists, social scientists, and the user communities. For these reasons, there exists an opportunity for a thorough assessment of the role of federally funded hurricane research in reducing societal vulnerability to hurricanes. The findings of such an assessment could contribute significantly to recent debates on the efficacy of federally funded research.

1.3.6 The challenge: toward a more usable science

Recent calls that scientific research show a closer connection to related societal benefits come at the same time as budget deficits limit the amount of federal resources designated for scientific research. These twin pressures on the research community increase demands upon science to demonstrate its use and value in addressing societal needs. In spite of policymakers' wishes, demonstration of the "use" of scientific information is in many instances neither straightforward nor simple. Traditionally, the scientific community has produced information with little, if any, serious or systematic consideration of its use or its users. The guiding principle for users, including sponsors of research, has generally been *caveat emptor* – let the buyer beware.

The general challenge facing US science policy is to better connect scientific research with societal benefits. In the context of the hurricane threat facing the US Gulf and Atlantic coasts, the challenge is to improve public and private decision-making with the aid of scientific research. With the modernization of the National Weather Service, a multi-billion dollar US Global Change Research Program, and a US Weather Research Program, weather and climate forecasts on various time-scales appear to be a growth industry. Yet, as Glantz (1986) has cautioned, "forecasts are the answer, but what was the question?"

CHAPTER 2

The US Hurricane Problem

2.1 REFRAMING THE US HURRICANE PROBLEM

In recent decades, damages from hurricanes have been rising rapidly in the United States (Figure 2.1). Recent examples include Hugo (1989, >$9 billion), Andrew (1992, >$30 billion), Opal (1995, >$3 billion), and Fran (1996, >$5 billion) (data, in current dollars, from the National Climatic Data Center and the National Hurricane Center). The rapid rise in hurricane-related damages has led many mistakenly to conclude that severe hurricanes have become more frequent in recent decades. For instance, a 1995 Senate report on federal disaster assistance asserted incorrectly that hurricanes "have become increasingly frequent and severe over the last four decades as climatic conditions have changed in the tropics" (BTFFDR 1995). In fact, the past several decades have seen a *decrease* in the frequency of severe storms and the period 1991–1994 was the quietest in at least 50 years (Landsea et al. 1996). Others have interpreted the trend of decreasing hurricane-related casualties to mean that hurricanes are no longer a serious threat. Consider the following lead to a Reuter's news article.

> Great killer hurricanes, like those seen in decades past, appear to be gone forever from the shores of the United States because of early warning systems. (Reuter's News Service 1996)

In contrast, in 1995 the director of the National Hurricane Center wrote that a "large loss of life is possible unless significant mitigation activities are undertaken" (Sheets 1995).

Taken together, the decrease in intense hurricanes coupled with the rapid rise in damages leads to a troubling conclusion: The United States is today more vulnerable to hurricane impacts than it has ever been – particularly if hurricanes again become more frequent (cf. AMS 1986).

2.1.1 The challenge of problem definition

Society's efforts to respond to hurricanes will be enhanced with a systematic understanding of the issues associated with hurricane impacts. Often, people neglect to identify the problem that they face, leading to misdirected solutions

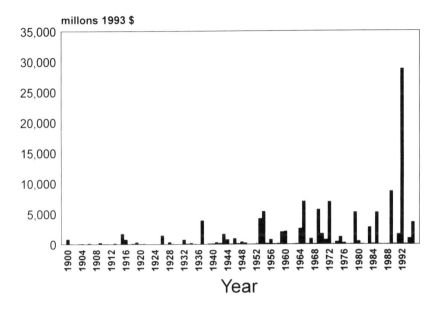

Figure 2.1   Annual hurricane damages, 1900–1995. Source: Hebert, Jarrell and Mayfield (1996)

with unintended consequences. If the hurricane problem faced by the United States is to be dealt with effectively, then an important first step is to understand the nature of that problem, and the implications of various alternative definitions of it.

Problems originate from the universe of "issues", which have been defined as "patterns of events with significance for human values" (Rein and White 1977). For instance, global climate change went from an esoteric scientific issue to an international problem when temperature trend data were associated with societal impacts of climate. Global climate change did not emerge as a policy problem overnight – observers will point out that climate change has been a topic of discussion in scientific circles for close to a century. Why did global climate change, or any other problem for that matter, emerge from the "policy primeval soup" to occupy a place on the public agenda (Kingdon 1984)? Why do problems emerge when they do? What role ought the social and physical science community play in shaping and responding to policy problems? Answers to questions like these lie in a deeper understanding of the role of problems and problem definitions in the policy process.

From issues to problems: getting on the agenda

The first step on the path from issue to problem is a sense of dissonance. John Dewey, the American philosopher, observed early this century that conscious human action is motivated by a "felt difficulty", that is, "a situation that is ambiguous, that presents a dilemma, that proposes alternatives. As long as our activity glides smoothly along from one thing to another . . . there is no need for reflection" (Dewey 1933). The step from issue to difficulty is an interpretive one – it is the perceptions of people that define which issues are considered important and which are not (Kingdon 1984). A perceived difficulty is not necessarily a problem, "a difficulty is only a problem if something can be done about it" (Wildavsky 1979). To understand or assess whether a particular difficulty is amenable to solution requires reflection, otherwise known as thinking, research, or inquiry. Through conscious thought, a person or a group is able to choose a course of action that they expect will improve their condition (this is called "rational" behavior) (Forester (1984) provides a useful review). But before an action can be chosen, alternatives must be available. In most cases the development of alternative courses of action depends upon how a problem is framed or defined.

People view the world through simplified "maps" or "models" that we create in our minds. Walter Lippmann (1961) referred to this as "the world outside and the pictures in our heads", and cognitive psychologists have explored the phenomenon in great detail. Definitions of problems are examples of such "maps" of the world. Such problem definitions allow for conscious reflection on ends to be sought (i.e. goals) and the means to achieve the desired ends. By definition, a problem is a "perceived discrepancy between goals and an actual or anticipated state of affairs" (Lasswell 1971). In other words, a problem is a difference between the way things seem to be and the way that we would like them to be (Glantz 1977). Thus, a problem definition contains (explicitly or implicitly) some sense of goals or objectives and some measure of (non)attainment with respect to those goals.

We use problem definitions to frame our social conditions. As a consequence, policy actions are directly tied to how social problems are framed. For example, if the problem of crime is defined as a consequence of the number of firearms available to criminals (a condition), then policy responses would likely focus on limiting or restricting firearm availability. Similarly, if the crime problem is defined as a consequence of a lack of education, then policy responses would likely emphasize a need for education. Because policy actions are so closely tied to problem definitions, it is important to pay close attention to how we define problems and not to allow problems to remain undefined or assumed.

Problem definitions can also blind us to aspects of the world that may be important to the invention, selection, and evaluation of alternative courses of action. Consider an example of a water resource controversy in Colorado (Bardwell 1991). Some people defined the problem as "we don't have enough water" leading to consideration of a range of alternative actions focused on "getting more water". However, others defined the problem as "we are using too much water" leading to consideration of a range of alternative actions centered on conservation and efficiency. The first definition of the problem blinded participants to a number of alternative response strategies to meet their water needs. How we define our problems often guides actions taken in response. A poorly defined or misdefined problem can lead to analytical "blind spots" (Stern 1986). Recall the story of the drunk who looked for his keys not where he dropped them, but under the streetlamp, because that was where the light was.

Because policy actions are highly dependent upon how societal problems are framed, a purpose of problem-oriented inquiry is to make explicit goals and measures of progress with respect to goals. Goal clarification – where we would like to be – is an often neglected but central task in the development of alternative actions to address problems. Goal clarification is itself an ongoing process that is intimately related to unfolding experiences. As political scientist E. S. Quade has noted:

> ... although there is widespread belief that goals should and can be set independently of the plans to attain them, there is overwhelming evidence that the more immediate objectives are – possibly more often than not – the result of opportunities that newly discovered or perceived alternatives offer rather than a source of such alternatives. (Quoted in Kaplan 1986)

Scientific research often provides insight into new opportunities and alternatives for actions to address problems (Mesthene 1967). For instance, with the development of the weather satellite, identification and location of hurricanes allowed for more effective warning and evacuation, thus changing the nature of the hurricane problem.

Of course, different people and groups define problems in different ways. The existence of different, often conflicting, problem definitions has political consequences. Even with the same information, value differences between individuals or groups often result in different conceptions of the existence, severity, or type of problem (Rein and White 1977). For instance, in the previously described water resources example, if on the one hand, the problem is a matter of water supply, then it emphasizes political power associated with control of water resources (Bardwell 1991). On the other hand, if the problem is a matter of conservation, the balance of political power shifts towards those with technology available to increase efficiency. Thus, different individuals often have different vested interests in particular problem definitions. In the public arena in the US, such differences are worked out through a process of bargaining, negotiation, and compromise under the provisions of the US Constitution.

Problem definition is further complicated by the existence of uncertain, imperfect, or partial scientific information (Etzioni 1985). Hence, various participants in a decision-making process will appeal to (and often selectively ignore) different scientific data for a host of reasons, e.g. to justify the primacy of their problem definition over others. Therefore, agreement on a problem definition by a broad range of participating individuals and groups facilitates efforts to act. Any analysis that recommends alternative actions to ameliorate a problem – such as lessons from a hurricane – will benefit from explicit definition of the problem to be addressed, including objectives to be achieved and how we might measure progress or lack thereof.

2.1.2 The conventional framing of the US hurricane problem

Conventional definitions of the hurricane problem facing the United States tend to focus on human casualties and property damage related to hurricane impacts. For example, the American Meteorological Society has stated that "the primary goal of both research and operational groups is to minimize loss of life from hurricanes" (AMS 1993). The Director of the National Hurricane Center testified before Congress: "we need action now to prepare for a return to more frequent major hurricane events in order to *minimize life and property loss*" (Sheets, 1994, emphasis added; cf. Sheets 1995; Simpson and Riehl 1981; Brinkmann 1975). The general goals of US hurricane policy that derive from the conventional definition of the hurricane problem are thus to minimize casualties and costs associated with hurricanes.

As a policy objective, "minimizing loss of life and property" is not a very useful way of framing the goal of US hurricane policy for the following two reasons (cf. Shabman 1994). First, the concept of "minimizing" sets *no casualties* and *no property losses* as the ultimate policy success. However, a full accounting of the costs and benefits of hurricane preparedness may find that elimination of all casualties and damages is too costly to achieve. In many cases calls for the minimization of loss of life and property damage often contain an implicit qualification "subject to the constraints of available funds, other resources, and technical and physical possibilities". These are important considerations that improve upon this particular problem definition, but because of the tendency to focus on the objectives of minimization of economic and human losses the qualifications are not always explicitly considered in practice.

A second reason why "minimizing loss of life and property" is not a particularly useful manner in which to frame the goal of US hurricane policy is that it is not clear what "minimization" means in practice: in the language of economics it is an "optimizing problem" that involves "multi-objective planning" and is "the way the objective function of the hurricane problem must be framed" (see Pielke (1997)). However, framing the goal of US hurricane policy in this manner reduces it to a largely academic exercise and places conceptual obstacles between the definition of the problem and the conditions which it represents, especially for the vast majority of people who deal with the hurricane-related issues on a day-to-day basis. These people often do not have expertise in "optimizing problems" or "objective functions".

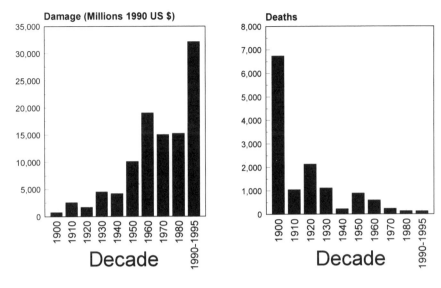

Figure 2.2   Trends in property damage and loss of life due to hurricanes by decade.
Source: Hebert, Jarrell and Mayfield (1996)

As a consequence, the conventional definition of the hurricane problem can mislead (and blind) us to various alternative actions. For instance, if one examines trends in loss of life and property damage (Figure 2.2), one might conclude (as implied in the Reuter article discussed in the introduction to this chapter) that the hurricane problem is largely a matter of damage rather than loss of life because of the steady decrease in hurricane-related casualties in this century and the corresponding increase in damages. However, this trend is misleading because the *potential* for large loss of life remains significant (cf. Hebert, Jarrell and Mayfield 1993). Therefore, the hurricane problem might be better defined in terms of *potential* loss of life and *potential* costs related to hurricanes. However, a difficulty arises in that the assessment of the potential impacts of hurricanes on society involves a wide range of confounding factors, many of which defy easy analysis.

Clearly, one advantage of the conventional framing of the hurricane problem in terms of lives lost and damage incurred is that such concepts are relatively easily quantified and measured. However, it is possible to approach the hurricane problem from a perspective that places loss of life and costs in a broader perspective. If, rather than asking: "How can we minimize the loss of lives and damage to property related to hurricane impacts?", we ask: "What actions can we take to reduce our *vulnerability* to hurricanes?", it might be possible to discover important, but neglected or overlooked, avenues of action that have been outside the illumination of the conventional problem definition.

2.1.3 Societal vulnerability as an alternative framing of the hurricane problem

Once it was thought that a society's vulnerability to extreme events was simply a function of an event's characteristics and incidence in places where people were at risk. This led to conclusions that a prediction of an extreme event, coupled with a technological or engineering solution (e.g. a levee system to prevent floods or an accurate forecast of a pending severe storm), would be sufficient to reduce vulnerability (Anderson 1995).[1] However, as scientific and technical tools and techniques have advanced, so too have losses of human lives and property because of the impacts of extreme events (White 1994). The implication of the twin trends of increasing technical sophistication and increasing losses signaled that vulnerability had more to it than the "natural-hazard-causing-societal-losses" perspective implied. There were obviously other factors at work.

A more accurate conception of societal vulnerability emphasizes the role that people play in creating their own vulnerabilities, as well as the role of others. A number of factors are responsible for human-caused vulnerability:

> Whereas previous assessments focused on acts of nature that come from outside human agency, later assessments acknowledged that it is largely human actions, decisions, and choices that result in people's vulnerability to natural events. Choices about where to live (or, in some cases, the lack of choice due to political, economic, or social position), decisions about where to locate a chemical plant, and acts of cutting forests, farming marginal lands, or evading building codes are examples of how humans cause a "natural" hazard to become a disaster. Humans make themselves – or, quite often others – vulnerable. (Anderson 1995)

Vulnerability is a concept from the literature on epidemiology, economics, and risk (Downing 1992) that has been used to define climate-change-related problems associated with hunger (e.g. Downing 1991), agriculture (e.g. Yarnall 1994), and sea-level rise (e.g. UNEP 1989). Vulnerability as a tool for climate-change-related problem definition has been advanced as an alternative to global circulation and economic model-driven scenarios of future socio-economic and climatic conditions.

> This shift of analytic perspective emphasizes the importance of vulnerability as a factor that amplifies or mitigates the impacts of climate change and channels them towards certain groups, certain institutions and certain places. It also emphasizes the degree to which the risks of climate catastrophe can be cushioned or ameliorated by adaptive actions that are – or can be brought – within the reach of populations at risk. (Downing 1991)

---

[1] An early work on vulnerability in the context of climate is Timmerman (1981). In the area of natural hazards more broadly see Hewitt (1983), Palm (1990), Alexander (1991), Winchester (1992), Kates et al. (1993), and Blaikie et al. (1994).

Vulnerability analysis allows for *expansion* of the degrees of action by allowing for the "possibility that human actions are independent variables, not just responses to climate change" (Downing 1991). In other words, definition of the climate change problem in terms of vulnerability to climate impacts may shed light on aspects of the issue that are neglected by conventional problem definitions (cf. Brunner 1991; Glantz 1988).

In the concept of "vulnerability" the analyst seeks to capture an "aggregate measure of human welfare that integrates environmental, social, economic, and political exposure to a range of potential harmful perturbations" (Bohle, Downing and Watts 1994). Chambers (1989) further refines the concept to "two sides". There is an "external side of risks, shocks and stress . . . and an internal side which is defenselessness, meaning a lack of means to cope without damaging loss". Vulnerability to a climate- or weather-related event is a function of society's exposure and of event incidence. Kates (1980) put it more eloquently: "the impacts of natural events are joint outcomes of the state of nature and the nature of society".

In defining a problem in terms of vulnerability, the general goal becomes the reduction of vulnerability. Of course, specific actions taken to reduce vulnerability in a particular context will be dependent upon how vulnerability is defined and the realism, cost, and practicality of alternative response strategies. The goal of "reducing vulnerability" is useful if it is general enough to garner sufficient consensus of opinion to allow for legitimate debate over various means to achieve the goal.

The concept of vulnerability can be used to reframe the hurricane problem facing the United States. Conventional definitions of the hurricane problem tend to rely solely on its physical aspects, often expressed in terms of a "risk", or focus exclusively on the economic or human losses related to hurricane impacts. Both types of definition neglect important aspects of the hurricane problem, and thus could lead to "blind spots" in efforts to develop response strategies, potentially resulting in opportunities missed or perils unseen.

2.2 VULNERABILITY TO HURRICANES

The vulnerability of society to hurricanes is a function of event *incidence* and societal *exposure*. Clearly, if people were not exposed to hurricanes or if hurricanes did not occur (i.e. no incidence), then society would be invulnerable to hurricanes. Incidence refers to the climatology of hurricanes – how many, how strong, and where. Exposure refers to the number of people and amount of property threatened by hurricanes. The number of exposed people and amount of threatened property can be reduced through preparedness efforts such as insurance, evacuation, and building fortification, where the term preparedness refers in a broad manner to all mitigation, response, and recovery activities. Societal vulnerability, then, is determined through the societal and climatic aspects of the hurricane phenomenon.

2.2.1 Hurricane incidence

Incidence is a function of climatic variability, including intensity, occurrence, and landfall frequency. The focus of the following discussion is on intense (also called "major", Category 3, 4, 5) hurricanes.

Intensity

Under the atmospheric conditions favorable for hurricane intensification (see Section 3.1.2), there are generally two types of intense hurricanes that threaten the US Gulf and Atlantic coasts. The first type are slow-moving storms such as usually occur in the Gulf of Mexico and over Florida and the southeast coasts. The slow movement of these storms over warm ocean waters means that winds in the same direction as the storm's movement will increase on the order of about only 10%. The second type of intense storm is one that has intensified over a warm sea surface and then is rapidly ejected to higher latitudes. In this scenario the rapid transit over cooler ocean waters does not permit much weakening of the storm prior to landfall. These storms threaten the Northeast Atlantic coast with high winds as a consequence of the storm's rapid forward motion. In the second type of storm, the area of maximum winds remains offshore in storms that move parallel to the Atlantic coast (with the eye remaining offshore).

The intensity of an Atlantic or Gulf of Mexico hurricane at landfall is directly related to its central pressure (Anthes 1982; Simpson and Riehl 1981) and to its speed. The potential minimum central pressure is limited by sea surface temperature (Merrill 1985; 1987): the higher the sea surface temperature, the lower the potential minimum central pressure and the stronger the storm. For example, the minimal Category 3 storm (which corresponds to a potential minimum central pressure of 964 mb) requires a sea surface temperature of at least 26.4°C (Merrill 1985). Hurricane Gilbert (1988), a Category 5 hurricane, is the most intense Atlantic hurricane on record, with a central pressure of 888 mb (Willoughby et al. 1989). Since 1899, only two Category 5 hurricanes have made landfall on the US coast, Camille in 1969 and an unnamed Labor Day storm that struck the Florida Keys in 1935 (Hebert, Jarrell and Mayfield 1993). Hurricane Opal (1995) was at Category 5 strength immediately prior to its landfall as a Category 3 storm.

Damage potential due to strong winds is directly related to a storm's central pressure and speed of forward movement. The speed of a storm is added (approximately) onto the wind speeds determined by the central pressure. For instance, the New England Hurricane of 1938 was traveling at over 70 knots at landfall (Coch 1994). Thus, winds parallel and in the direction of the 1938 storm's direction of motion were increased substantially. The 1938 storm is an example of a hurricane that has intensified over a warm sea and then ejected to a higher latitude.

Occurrence

Intense hurricanes are irregular events that occur with greater and lesser annual frequencies over decadal and longer time-scales, at particular locations on the coast. For example, from 1930 to 1959, 25 intense hurricanes struck the US; from 1960 to 1989 only 15 made landfall. Figure 2.3 shows intense hurricane tracks for the periods 1947–1969 and 1970–1987. In the case of Florida, decadal variability in hurricane incidence is striking. Figure 2.4 shows hurricane and tropical storm landfall for Florida for the periods 1941–1950 and 1971–1980. In the earlier decade 20 storms made landfall (six intense); in the later decade there were only four (one intense). When intense hurricanes do occur, they are intensively studied (e.g. Hurricane Frederic, Powell 1982; Hurricane Hugo, Golden 1990; Hurricane Andrew, Wakimoto and Black 1994). Although the hurricane season lasts from June to November, 95% of Category 3, 4, and 5 hurricanes in the Atlantic Basin occur from August to October (Landsea 1993).

Since 1984, William Gray of Colorado State University and a team of researchers have predicted the following season's hurricane activity based on a number of documented statistical relationships (see Section 4.5.3). His Atlantic Seasonal Hurricane Variability technique is based on several climatic indices (Gray 1994). Among the statistical forecasts made using this method are Hurricane Destruction Potential, Intense Hurricanes (reaching at least level 3 at some time in a storm's evolution), and Intense Hurricane Days. Gray's forecasts have received wide attention in the media and by user communities (e.g. Morgenthaler 1994).

It has been suggested by some scientists that the hurricane season may be forecast with longer than a year lead time with some level of forecast skill (Landsea et al. 1994), including speculation regarding a connection between decadal variations in the "ocean conveyor belt" and sea surface temperatures, Sahel rainfall, El Niño–Southern Oscillation (ENSO) cycle, and Atlantic hurricanes. Gray (1992) stated that:

> . . . of those 24 hurricane seasons of this century in which major hurricane landfall occurred [along the US East Coast] in only one season (1992) did distinctly dry Western Sahel rainfall conditions exist in the same year with distinctly warm El Niño sea surface conditions. And, that year and storm was 1992 and Andrew.

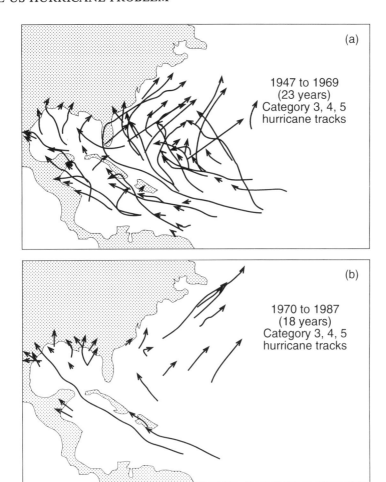

Figure 2.3 Comparison of intense (Saffir/Simpson Categories 3, 4, and 5) hurricane tracks for (a) 1947–1969, and (b) 1970–1987. Since 1989 there have been five landfalling intense hurricanes, Hugo (1989) Andrew (1992), Emily (1993), Opal (1995), and Fran (1996). Graph provided by W. Gray

The occurrence of Category 3, 4, and 5 hurricanes along the US East Coast has a strong correlation with Sahel rainfall (Gray 1990; Gray and Landsea 1992; Goldenberg and Shapiro 1996). With a reliable prediction of the ENSO cycle or western Sahel rainfall more than one year ahead, it may be possible also to forecast reliably interannual hurricane occurrence. At present, however, such skills are in their infancy.

Based on the work of Gray and others, there appears to be some statistical skill in predicting the level of hurricane activity in the Atlantic ocean, at least

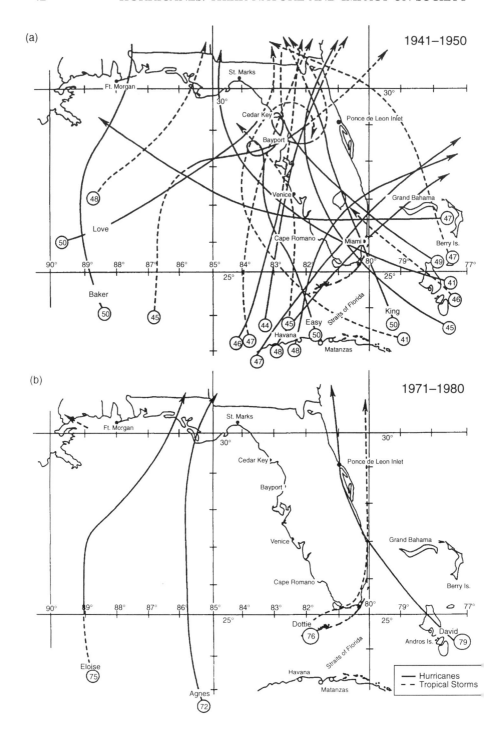

on an intra-annual basis, and with some suggestion that longer period skill may be achievable (see Section 4.5.3 for discussion). However, none of these techniques offers a tool for predicting with skill specific landfall locations at annual or longer time-scales (Hess and Elsner 1994). Thus, the historical climatological record remains the best tool available to estimate coastal vulnerability to hurricane occurrence for specific locations. In other words, to assess hurricane occurrence for a particular location it is better to look back than to look ahead.

Landfall frequency

Over the past several decades there has been a decrease in major landfalling hurricanes (Landsea et al. 1996). Hurricanes Hugo, Andrew, and the active 1995 and 1996 hurricane seasons have led many to suggest that annual hurricane activity is increasing. Over the past century, a Category 4 storm strikes the US, on average, every six years (Hebert, Jarrell and Mayfield 1996). Hurricanes Andrew and Hugo are the only Category 4 storms to make landfall on the US coast since 1969, and Andrew in 1992 was the first intense hurricane to make landfall in South Florida since Betsy in 1965.

On average, two hurricanes strike the US coast each year, with *two intense hurricanes striking the US coast every three years* (Hebert, Jarrell and Mayfield 1996). Therefore, if future rates recur at their historical average, the average damage due to an intense hurricane would only need to be $4.5 billion for annual US exposure to be at least $3 billion. Hurricanes Hugo, Andrew, Opal, and Fran suggest that this estimate may be low. If the average damage due to an intense hurricane is $7.5 billion, for example, then annual exposure would be at least $5 billion. Such estimates exclude the costs of landfalling tropical storms and Category 1 and 2 hurricanes, which are capable of extensive economic damage (cf. Landsea 1991). For example, tropical storm Gordon (1994) resulted in more than $200 million in agriculture-related damage in Florida (NYT 1994).

Other statistical information related to landfall frequency includes: (a) more than a third of all US landfalling hurricanes hit Florida; (b) more than 70% of US landfalling hurricanes of Category 4 or 5 hit Florida and Texas; and (c) along the middle Gulf Coast, southern Florida, and southern New England half of all landfalling hurricanes are Category 3 and higher.

Figure 2.5 shows the probability of a direct hit by an intense hurricane, based on historical data, expressed as an annual percentage for each of the 168 coastal counties from Texas to Maine (see also Section 2.2.3 and

Figure 2.4   Hurricane and tropical storm tracks for storms which made landfall over Florida for (a) 1941–1950, and (b) 1971–1980. Source: Doehring, Duedall and Williams (1994)

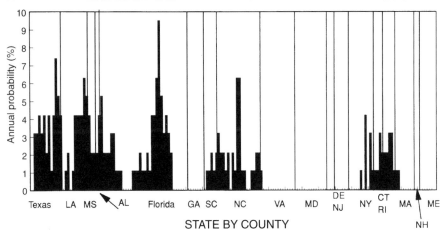

Figure 2.5   Observed (1900–1994) annual probability of a direct hit by an intense hurricane for US coastal counties from Texas to Maine (updated from Jarrell, Hebert and Mayfield 1992)

Figure 2.7).[1] The highest probability is for Monroe County, Florida (0.095), immediately to the south of Dade County where Hurricane Andrew (1992) made landfall. The Gulf Coast and the Atlantic Coast of Florida are particularly at risk to hurricane landfall. In contrast, several counties along the Gulf Coast and many along the Atlantic Coast have never experienced a direct hit from an intense hurricane during the 20th century. Most of New England sees few direct hits from hurricanes because, in general, storms accelerate northward or northeast on a track that typically places the storm center either inland or parallel to the coast but offshore. The absence of direct hits to the northern Florida and Georgia coasts is partly a consequence of their orientation with respect to the more typical hurricane track as the storms begin to recurve around the subtropical Bermuda high-pressure ridge that has a clockwise wind circulation, centered over the mid-Atlantic Ocean. Undoubtedly, the absence of direct hits is also partly due to good fortune (cf. Kocin and Keller 1991). A landfall of Hurricane Hugo (1989) just slightly further south would have altered these statistics. Indeed, a number of hurricanes struck the coast of northern Florida and Georgia in the 18th and 19th centuries. For instance, the Atlantic Coast between St Augustine, Florida, and Savannah, Georgia, was hit by very strong storms in 1824 and 1837 (Ludlam 1963).

---

[1] Jarrell, Hebert and Mayfield (1992) compile hurricane landfalls in coastal counties since 1900 and provide the data necessary to compute landfall probabilities. One storm which could arguably be added to their compilation is the 17–26 August 1933 storm that made landfall at the North Carolina Outer Banks with a central pressure of 960 mb and moved north along the western side of the Chesapeake Bay (Cobb 1991).

Table 2.1 shows annual landfall probabilities (as percentages) for intense hurricanes, and year of last direct hit by an intense hurricane for selected coastal counties and major metropolitan areas for each state from Texas to Maine, and also displays for each coastal county insured property values and estimated 1993 population: An observed annual landfall probability of 0.053 for Miami, Mobile, and Galveston corresponds to a greater than 50% chance of at least one intense hurricane within the next 13 years. The calculation which gives the period N, where N equals the number of years from now within which there exists a greater then 50% chance of at least one intense hurricane, is $(1-P)^N \geq 0.050$, where P is the annual landfall probability based on the historical record, is assumed to be constant and events are independent from year to year. Similarly, an observed annual landfall probability of 0.021, such as observed in Charleston, South Carolina, Wilmington, Delaware, and Providence, Rhode Island, corresponds to a greater than 50% chance of at least one intense hurricane within the next 33 years. For Monroe County, Florida, immediately to the south of Dade County (Miami), there is a greater than 50% chance of at least one intense hurricane in the next seven years.

The historical record represents a baseline against which efforts to improve long-term forecasts ought to be measured. Skillful forecasts that improve upon the landfall probabilities generated from the historical record may have value to those decision-makers who can alter their behavior accordingly.

The probabilities shown in Figure 2.5 for intense hurricanes are similar to the findings of Ho et al. (1987) for all types of landfalling hurricanes. They find that:

Highest frequency of landfalling tropical cyclones on the east coast is in southern Florida, and a comparatively high frequency appears to the south of Cape Hatteras, North Carolina. The frequency of entries drops off rapidly from Miami to Daytona Beach, Florida and from Cape Hatteras northward to Maine, except around Long Island.

Figure 2.6 shows analyses of landfall probabilities from two studies for purposes of comparison (Figure 2.6a: Simpson and Lawrence 1971; Figure 2.6b: Ho et al. 1987). Simpson and Lawrence (1971) report observed landfall probability for 50-mile segments of coast. Differences between Simpson and Lawrence (1971), Ho et al. (1987) and the analysis of landfall probability in this chapter are primarily due to different levels of analysis – that is, landfall probability percentages are given in this chapter for each coastal county, whereas Simpson and Lawrence (1971) and Ho et al. (1987) give landfall probabilities for equal segments of coastline. Each method has its strengths: equal segments of coastline allow for ready comparison between segments, whereas data by county may be more meaningful to decision-makers at state and local levels.

Table 2.1 Summary of hurricane statistics for selected locales in Gulf and Atlantic coastal states

State	City	County	Population 1993 (est)* (000s)	Insured property 1993[†] (billions $)	Observed intense landfall probability (%/year)	Most recent intense direct hit (year, category, storm)[§]
TX	Brownsville	Cameron	270	13.182	3.2	1980 (3) Allen
	Galveston	Galveston	217	19.309	5.3	1983 (3) Alicia
LA	New Orleans	Orleans	491	43.340	4.2	1965 (3) Betsy
MS	Gulfport	Harrison	166	13.470	2.1	1985 (3) Elena
AL	Mobile	Mobile	384	28.606	5.3	1985 (3) Elena
FL	Tampa	Hillsborough	847	69.968	1.1	1921 (3)
	St Petersburg	Pinellas	859	71.283	1.1	1921 (3)
	Miami	Dade	1979	160.844	5.3	1992 (4) Andrew
	Jacksonville	Duval	689	55.527	0.0	1854
GA	Savannah	Chatham	219	22.386	0.0	1893
SC	Charleston	Charleston	304	24.118	2.1	1989 (4) Hugo
NC	Wilmington	New Hanover	125	11.814	2.1	1996 (3) Fran
VA	Norfolk	Norfolk	261	18.912	0.0	1856
MD	Ocean City	Worcester	35	6.269	0.0	1933 (3)[∞]
	Baltimore	Baltimore"	1438	117.128	0.0	1850
DE	Wilmington	New Castle	450	49.794	0.0	1861
NJ	Asbury Park	Monmouth	227	62.038	0.0	1861
NY	New York City	Kings	2289	124.887	0.0	1821
CT	New Haven	New Haven	803	87.142	1.1	1938 (3)
RI	Providence	Providence	596	45.305	2.1	1954 (3) Carol
MA	Cape Cod	Barnstable	187	29.572	1.1	1954 (3) Edna
	Boston	Suffolk	649	74.299	0.0	1869
NH	Portsmouth	Rockingham	243	28.010	0.0	1788
ME	Portland	Cumberland	245	23.341	0.0	1830

* Source: US Census Bureau; [†] Source: Insurance Institute for Property Loss Reduction; [‡] Source: Jarrell, Hebert and Mayfield 1992. Period of record is 1900–1994; [§] Data for years since 1900 from Hebert, Jarrell and Mayfield 1996. Estimates for years prior to 1900 are based on descriptions in Ludlam (1963) and Tannehill (1952); [∞] According to data in Cobb (1991) the 1933 storm was a Category 3 hurricane. " Includes both city and county.

46

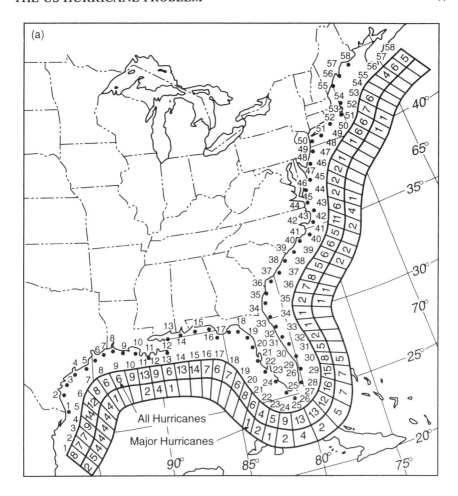

Figure 2.6   Analyses of landfall probabilities from (a) Simpson and Lawrence (1971) and (b) Ho et al. (1987) for purposes of comparison

2.2.2 Hurricane incidence and climate change

It has been suggested that hurricane intensity, occurrence, and landfall frequency may be affected by human-caused global warming. One hypothesis is that the oceans would warm, thereby creating the potential for more intense hurricanes (Emanuel 1987) (although Gray (1990) argues that variability in hurricane incidence and intensity is a function of natural climatic variability). Based on such scientific hypotheses, a number of groups have argued that the recent impacts of Hurricanes Andrew and Hugo are evidence for global warming (e.g. Leggett 1994).

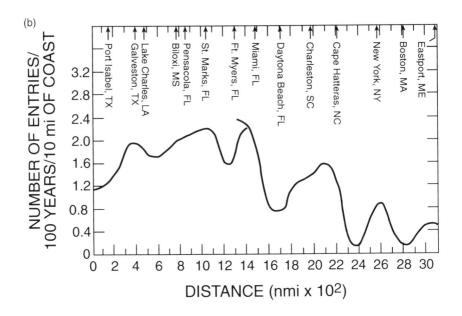

Figure 2.6 (*continued*)

According to a group of scientists convened to report on the possibility of climate change, the historical record offers no confirmation of either claim.

> Overall there is no evidence that extreme weather events, or climate variability, has increased, in a global sense, through the 20th Century, although data and analyses are poor and not comprehensive. On regional scales there is clear evidence of changes in some extremes and climate variability indicator. Some of these changes have been toward greater variability, some toward lower variability . . . it is not possible to say whether the frequency, area of occurrence, time of occurrence, mean intensity or maximum intensity of tropical cyclones will change. (IPCC 1996a)

For instance, intense hurricanes have been decreasing in the Atlantic over the past 50 years (Landsea et al. 1996), and typhoons have first decreased, then increased in the western North Pacific (Chan and Shi 1996). In short, while science today provides little guidance as to exact future hurricane incidences, the historical record does provide a sense of what might be expected.

The global warming debate will likely continue for some time, including arguments over future hurricane incidences. However, with certainty it can be said that in addition to faulty logic (i.e. the converse of a true proposition is not necessarily true) such arguments, no matter how well intentioned, serve to direct attention away from the documented hurricane threat. These arguments focus attention on responses to global warming (e.g. reducing carbon emissions) as a means to address society's hurricane problems, rather than on the need for increased hurricane preparedness in individual communities to reduce vulnerability.

Even with the uncertainty expressed by the scientific community, a prevailing view is that increased hurricane impacts can be prevented through policies in response to global warming. For instance, the cover of the 16 January 1996 *Newsweek* had the following statement: THE HOT ZONE: HURRICANES, FLOODS, AND BLIZZARDS, BLAME GLOBAL WARMING. The implication of the statement is based on the following logic. If global warming can be prevented then increased impacts associated with hurricanes can also be prevented. However, the logic is incomplete, and thus misleading, in that it neglects an important lesson of experience. First, in the United States at least, hurricane losses have increased dramatically during a period of *decreasing* intense hurricane incidence. Given that the trend in impacts is almost entirely due to changes in society, it is reasonable to expect that impacts will continue to increase dramatically, unless actions are taken to reduce vulnerability, and this need is largely independent of the number of future storms (Pielke and Landsea 1997). In the context of hurricanes, measures to reduce vulnerability make sense under any scenario of change in climate, i.e. increasing, or constant levels of hurricane incidence.

Thus, the issue of global warming is largely irrelevant to the need for actions to reduce our vulnerability to hurricanes: history alone dictates that such actions are sorely needed.

2.2.3 Exposure to hurricanes

Exposure to hurricanes is a function of (a) population at risk, (b) property at risk (cf. Brinkmann 1975), and (c) preparedness. This section presents data on hurricane exposure at the coastal county level for 168 coastal counties that lie adjacent to the Gulf of Mexico and the Atlantic Ocean from the Mexico–Texas border to the Maine–Canada border. Figure 2.7 shows the coastal counties. Of course, societal vulnerability to hurricanes extends well inland, beyond the coastal counties. For instance, following Hurricane Andrew's landfall in Louisiana, 29 inland parishes were declared disaster areas in addition to all 11 of Louisiana's coastal parishes (DOC 1993). The remnants of Hurricane Camille (1969) killed more than 100 people in central Virginia as a result of up to 30 inches (0.75 m) of rain from floods within six hours. Coastal counties, however, are a primary component of societal vulnerability to hurricanes.

Population at risk

Figures 2.8a–d show US coastal county population by state for each of the 168 coastal counties from Texas to Maine for the years 1930, 1950, 1970, and

Figure 2.7   The 168 coastal counties from Texas to Maine used in this study

1990 (US Census). For purposes of comparison, Figure 2.9 shows the four graphs on one page. Figure 2.10 shows projected coastal county populations for the year 2000 based on US Bureau of the Census state level projections (Campbell 1994).

The most readily apparent trend is the growth in population along the Gulf and Atlantic counties from Texas through North Carolina. For instance, in 1990 the combined population of Dade County, Florida, and its two neighbors to the north, Broward and Palm Beach Counties, more than four million people, was greater than that of 29 states. About the same number of people lived in Dade and Broward counties in 1995 – 3.2 million – as lived in *all* of the 109 coastal counties from Texas through Virginia in 1930. A second trend is the very low level of growth in the coastal counties of the northeast US. Some counties north of New York City have actually experienced popu- lation decreases in recent decades. However, the population of the Atlantic Coast from Baltimore to Boston remains very large.

Rising population has not only created record densities for areas such as south Florida, but also "filled in" formerly low-population areas. In the 95

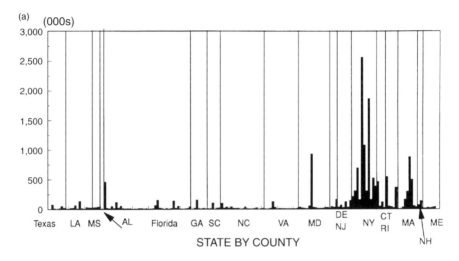

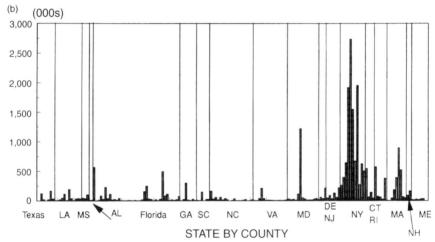

Figure 2.8   US coastal county population by county and state for (a) 1930; (b) 1950; (c) 1970; (d) 1990. The vertical lines in these figures represent state boundaries with states listed along the horizontal axis. Source: US Census Bureau

coastal counties from Texas through North Carolina the number of counties with populations of more than 250 000 tripled from 1950 to 1970, from three to nine, and doubled again from nine to 18 by 1990. A quarter of a million residents is about the population of Charleston County, South Carolina, where Hurricane Hugo made landfall in 1989 and caused $8.2 billion (1993 $) in damage (Hebert, Jarrell and Mayfield 1993). Hurricane Hugo, however, made landfall in a relatively unpopulated stretch of South Carolina coast (Baker 1994). Had Hugo directly hit a more populated section, casualties and

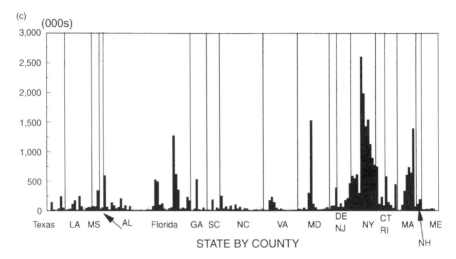

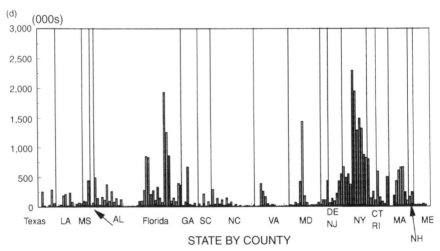

Figure 2.8   (*continued*)

damage would likely have been significantly higher. The number of counties
with more than 100 000 residents went from 15 in 1950 to 21 in 1970, and to
36 in 1990. Hurricane Frederic made landfall in a county of about 100 000
near Gulf Shores, Alabama, in 1979 and caused $3.8 billion (1993 $) in
damage (Hebert, Jarrell and Mayfield 1993). Another way to look at population growth is in terms of the dwindling number of counties with very few
residents. From Texas through North Carolina, the number of counties with
fewer than 50 000 residents decreased from 75 in 1950 to 54 in 1970, and to
38 by 1990. Hurricane Andrew made landfall across two such relatively low-
population counties in Louisiana in 1992 and caused about $1 billion (1993 $)

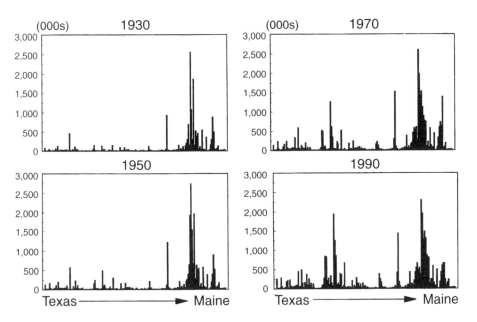

Figure 2.9   Comparison of coastal county populations by county and state for 1930, 1950, 1970, and 1990

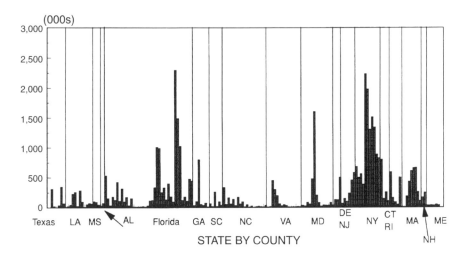

Figure 2.10   Projected US coastal county population by county and state for 2000. Source: US Census Bureau

in damage (Cochran and Levitan 1994). As Figure 2.10 shows, population growth is expected to continue in most coastal counties. In aggregate, in 1990 the coastal counties were home to more than 40 million people, or about 16% of the total US population. Although the numbers are large, census populations may actually underestimate the number of people along the coast during hurricane season. Because the hurricane season overlaps with the tourist season in many of these coastal counties, many more people in addition to permanent residents may actually be in the path of an approaching hurricane (cf. Sheets 1993).

Property at risk

Figure 2.11 shows insured property values in each of the 168 coastal counties from Texas to Maine for the years 1988 and 1993, based upon a study by the Applied Insurance Research in collaboration with the Insurance Institute for Property Loss Reduction. Figure 2.12 shows the increase from 1988 to 1993 as a percentage of the 1988 total. An increase of 100% represents a doubling in value, 200% a tripling, etc. Inflation accounts for 19.5% of the aggregate growth during that five-year period (Council of Economic Advisors 1994). The remainder of the growth can be attributed to expanded insurance coverage and real increases in property values.

The total amounts of insured property are staggering. Over $3.1 *trillion* worth of property was insured in 1993, an increase of 69% (50% excluding inflation) over the 1988 total of about $1.9 trillion. The 1988 total represented an increase of 64% (35% after inflation) over the 1980 total of about $1.1 trillion (Sheets 1993). For comparison, the coastal counties represented about 15% of the total insured property in the United States in 1993, which was about $21.4 trillion; for 1988 the figures are approximately 14% and $13 trillion, respectively.

As compared to the rapid coastal population growth in recent years, the growth in insured property in the five years from 1988 to 1993 is even more startling. After adjusting for the effects of inflation, the aggregate growth for all coastal counties is 46%, more than twice the inflation of the period. Table 2.2 shows a summary by state of insured property for the coastal counties. Except for Louisiana, Florida had the slowest rate of growth in coastal insured property. Despite its lower rate of growth, the total insured coastal property in Florida exceeds the combined coastal insured property from Texas to Delaware (excluding Florida). Within states, local variations are large.

Only one county, Laforche Parish, Louisiana, had property values decrease over the period (compensating for inflation). Over the five-year period, 24 counties experienced more than a doubling in insured property. Although the amount of insured property is large, insured property represents only a portion of the total potential losses due to hurricanes. Uninsured property and public infrastructure make up a substantial portion of potential damage due to

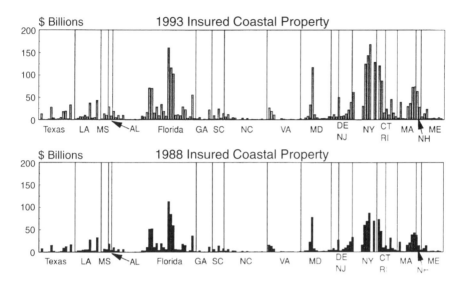

Figure 2.11   Insured US coastal county property values by county and state for 1988 and 1993. Data provided courtesy of Insurance Institute for Property Loss Reduction

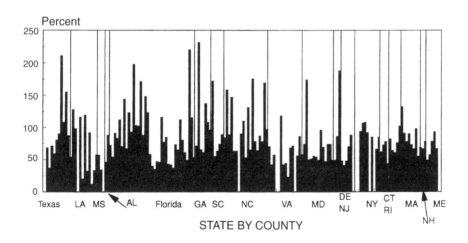

Figure 2.12   Increase in insured US coastal county property values by county and state from 1988 to 1993 as a percentage of the 1988 total. Inflation accounts for about 19.5% of the change over the period. Data provided courtesy of Insurance Institute for Property Loss Reduction

Table 2.2   Summary of coastal county insured property by state

State	Total coastal insured property values (current $ billions)		1988–1993 increase as a percentage of 1988 value
	1988	1993	
Texas	70.1	128.6	83
Louisiana	87.5	123.5	41
Mississippi	14.1	25.5	80
Alabama	22.8	36.9	61
Florida	565.8	871.7	54
Georgia	16.5	32.5	96
South Carolina	31.2	54.7	75
North Carolina	22.7	45.0	97
Virginia	42.5	67.8	59
Maryland	129.2	202.6	56
Delaware	38.7	67.7	74
New Jersey	88.5	152.8	72
New York	301.7	595.6	97
Connecticut	143.3	248.1	73
Rhode Island	52.9	83.1	57
Massachusetts	179.8	321.6	78
New Hampshire	18.5	34.9	88
Maine	32.3	54.5	68
Coastal total	1 858.1	3 147.0	69
US total	12 967.1	21 422.0	65

Source: Insurance Institute for Property Loss Reduction.

hurricanes. It is also important to recognize that the "costs" of hurricane impacts go well beyond those which can be expressed in dollars (e.g. Mauro 1992).

Preparedness

Throughout this discussion, the term "preparedness" is used in a general sense to refer to the full range of anticipatory and emergency management activities (e.g. mitigation, preparedness, response, restoration). Preparedness refers to all of the efforts at various levels of public and private decision-making to reduce hurricane-related casualties and damage (BTFFDR 1995; Wolensky and Wolensky 1990). One study argues that preparedness planning "makes excellent sense to do now those things which can reduce or minimize the risks and costs of future hurricanes, and hasten sensible recovery practices after the storm (Salmon and Henningson 1987). Preparedness has technical, practical, and political aspects which, in large part, are often determined by the idiosyncrasies of and resources available to each community. Therefore, levels of preparedness (and consequently, societal vulnerability) vary a great deal between communities along the US Gulf and Atlantic coasts.

In general, preparedness activities have short- and long-term components (Salmon and Henningson 1987). Short-term responses focus on a particular approaching storm. Long-term responses focus on the hurricane threat more generally. Many short-term responses related to protection of the exposed population are based upon long-term studies of expected storm surges due to landfalling hurricanes (Sheets 1990). For example, expected coastal flooding is calculated using the SLOSH (Sea, Lake, and Overland Surges from Hurricanes) storm-surge model, which is discussed further in Section 5.2.2. The SLOSH model was conceived, developed, and first applied by Chester Jelesnianski. The SLOSH model is run for 31 "SLOSH basins" along the US coast from Texas to Maine (Jarvinen and Lawrence 1985). Errors in the forecasts of specific landfalling storms are compensated for because each map represents a composite of maximum storm surges for a range of landfall points, storm movement, and intensity. Evacuation studies are then based upon the areas identified to be at risk from the output of the SLOSH modeling process. Such studies include a behavioral component that seeks to identify "realistic assumptions of how the public will behave when advised or ordered to evacuate" and a transportation analysis which seeks to identify the capacity of routes of escape, points of congestion (Baker 1993b; Carter 1993), and places of "last resort refuge" (Sheets 1992). Short-term response is focused on the hurricane forecast. The National Weather Service, National Hurricane Center, local and state officials, and the media coordinate hurricane watches and warnings based upon the forecast tracks of specific approaching hurricanes (Sheets 1990).

The following isolated incident, which occurred in South Carolina during Hurricane Hugo (1989), illustrates the stakes involved with long-term planning for short-term response.

In the village of McClellanville, the Lincoln High School was used as an evacuation shelter. The evacuation plan listed the base elevation of the school as 20.53 feet National Geodetic Vertical Datum (NGVD). Many of the residents took shelter in this school. During the height of the storm, water rose outside the school and eventually broke through one of the doors. Water rushed in and continued to rise inside the school reaching a depth of 6 feet within the building. A resident with a videocassette recorder documented people climbing on tables and bleachers to escape the rising water. As the water reached its maximum height, children were lifted onto the school's rafters. Fortunately, everyone survived the event although not without considerable anxiety.

Later examination revealed that the base elevation of the school was 10 feet, not the 20.53 feet listed on the evacuation plan. This school should not have been used as a shelter for any storm greater than a Category 1 hurricane. (DOC 1990)

Warning and response to Hurricane Hugo based upon the SLOSH model process have been generally judged successful (e.g. DOC 1990; Baker 1994;

Coch 1994), yet had the evacuees at Lincoln High been less fortunate such judgments would likely have been very different. The incident demonstrates the fine line between success and failure in long-term planning for short-term response in order to reduce vulnerability to hurricanes.

Another example is from Louisiana during Hurricane Andrew, when a number of emergency management officials had difficulty interpreting updated storm surge maps, and consequently relied on older, potentially dated information (USACE/FEMA 1993). Other officials in Louisiana did not have relevant software available to aid in the evacuation decision process (USACE/ FEMA 1993).

The protection of property at risk also has short-term and long-term components. Designing structures to withstand hurricane-force winds is an important factor in reducing property damage (e.g. Mehta, Cheshir and McDonald 1992). An essential aspect in the reduction of property damage is the establishment and enforcement of building codes commensurate with expected hurricane incidence. A 1989 study found that building code enforcement was a primary factor in reducing structural damage to buildings (Mulady 1994). One insurance official claimed that poor compliance with building codes accounted for about 25%, or about $4 billion, of the insured losses in south Florida due to Hurricane Andrew (Noonan 1993). Other estimates range upwards to 40%, or close to $6.5 billion.

Complacency is the enemy of preparedness. The *New York Times* reported in 1993 that "of 34 coastal areas identified as needing evacuation studies, less than half have been completed, and only $900,000 a year is available for commissioning new ones" (Applebome 1993). A FEMA official complained that the lack of "funding is inhibiting an aggressive and comprehensive approach to hurricane preparedness programming" (Applebome 1993). According to data provided by the Army Corps of Engineers in 1996, the situation remains much the same, with many incomplete and dated studies (R. Plott 1996, personal communication).

Complacency led to Dade County's lack of preparedness for Hurricane Andrew (Leen et al. 1992). For instance, in 1988 Dade county employed 16 building inspectors to serve a population of well over one million. On many occasions in the years preceding Andrew, inspectors reported conducting more than 70 inspections per day, a rate of one every six minutes, not counting driving time (Getter 1993). Such anecdotes highlight the need for systematic assessments of hurricane preparedness in the broader context of hurricane vulnerability *before a hurricane strikes*. There is sufficient evidence of complacency along the US Atlantic and Gulf coasts that future hurricane disasters should not come as surprises when they occur (cf. Sheets 1992; 1993; 1994; 1995).

With population-at-risk and property-at-risk in coastal areas rising rapidly, and the rate of growth increasing as well, the key to reduced hurricane exposure lies in improved preparedness. While past responses to reduce the threat to human life have been extremely successful, demographic changes mean that the nature of the hurricane threat is ever changing. The recent experience of Hurricane Andrew in Dade County, believed to be among the best prepared locales, suggests that many coastal areas may not be as prepared for hurricane impact as was once thought. Unless actions are taken to reduce vulnerability, the worst disasters lie ahead.

2.3 ASSESSMENT OF VULNERABILITY TO HURRICANES

Assessment of a community's societal vulnerability to hurricanes requires attention to a number of contextual factors. A short list includes demographic, political, policy, climatological, and behavioral factors. Vulnerability assessment is difficult because a community's vulnerability to hurricanes is constantly changing due to demographic and political shifts, as well as climatic variability. Furthermore, societal vulnerability is a difficult concept to measure (Timmerman 1981). However, through assessment of vulnerability it is possible to identify weaknesses, strengths and opportunities to improve the processes of decision-making.

While population and property at risk are easily estimated, a community's level of preparedness is more difficult to assess. However, efforts to reduce vulnerability are enhanced by the wealth of experience that society has with hurricane impacts. The structure of preparedness has largely been put in place in successive aftermaths of past events. Therefore, at a general level, preparedness is largely understood. One could easily argue that reduction of societal vulnerability to hurricanes is largely a matter of the *application* of existing knowledge, rather than of the *need* for fundamentally new knowledge.

2.3.1 Tropical cyclone risk assessment

An important aspect of the conventional definition of the hurricane problem is "risk" One way to assess risk is to anticipate the future. In order to plan for future hurricane impacts decision-makers use a variety of tools to obtain a sense of the potential scale of those future impacts. There are at least five different methods that decision-makers use to assess the risk of tropical cyclone impacts on people and property (Dlugolecki et al. 1996). The first three are based upon the documented historical record of tropical cyclone impacts.

- *Loss record*
  This bases loss estimates on experience, but its weakness is that "experience" is a relatively short record and may not be representative (cf. Wamsted 1993). It may not even be representative of climatological trends.

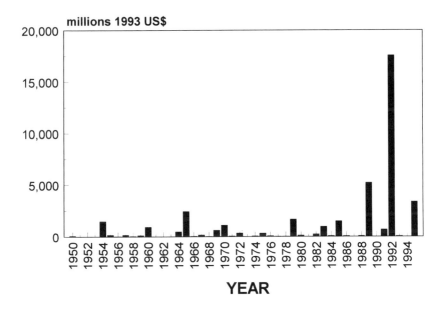

YEAR

Figure 2.13   Insured losses related to tropical cyclones in the United States, 1950–1995. Data provided courtesy of Property Claims Services, Inc.

Figure 2.13 shows data kept by Property Claims Services, Inc. of insured losses related to tropical cyclones in the US.

- *Event Models*
  The people or property subject to risk is inventoried based on a number of key dimensions (e.g. number, type, and location of structures) and then, based on the inventory, a computer model is created to estimate losses from a particular event's impact. A number of companies (such as Property Claims Services) run these models (Banham 1993). See Table 2.3 for an example.

- *Hypothetical Event Models*
  Modeling is used here as well, but the focus is not on a particular event but instead on a family of events and the corresponding frequency and magnitude distribution of impacts. Companies such as Applied Insurance Research and EQE International run these sorts of models (Banham 1993). See Table 2.4 for an example.

As is the case with all models, they are only as good as the assumptions which underlie them. For instance, prior to hurricane Andrew, models such as these led hurricane loss experts to conclude that the worst-case scenario for a

Table 2.3   Example of a specific event modeling approach

1992 Losses for past storms*

Name	Location	Date	1992 damages (US$ billions)
Cat. 4	Miami	1926	39
Andrew	S. Florida	1992	24
Betsy	S. Florida	1965	15
Donna	Sombrero Key	1960	10
Cat. 4	Pompano Beach	1947	9
Cat. 3	Homestead	1941	5
Cat. 4	Palm Beach	1928	3.5
Cleo	Miami	1964	2.7
Cat. 3	Palm Beach	1949	2.6
Inez	S. Florida	1966	2.2

Data from: Englehardt and Peng (1996).

* Damages are adjusted to 1992 estimates based on inflation and increased housing density.

Table 2.4   Example of a hypothetical event modeling approach

Hurricane loss projections for Category 4 or 5 hurricanes at key points along the coastline

Category	Location	Total insured loss (US$ billions 1993)
5	Miami, FL	52.5
5	Fort Lauderdale, FL	51.9
5	Galveston, TX	42.5
5	Hampton, VA	33.5
5	New Orleans, LA	25.6
4	Asbury Park, NJ	52.3
4	New York City, NY	45.0
4	Long Island, NY	40.8
4	Ocean City, MD	20.1

Source: Applied Insurance Research in IIPLR/IRC (1995).

hurricane impact along the US coast would be no more than $10 billion (e.g. Sheets 1992). Even in the immediate aftermath of Andrew, many estimates of damages were off by significant amounts (Noonan 1993). The primary reason for this was a number of important factors not included in the models that only became apparent in the wake of the disaster. Of course, for the insurance companies that sponsor such models, overall accuracy with respect to loss totals may not be the most important factor – relative impacts on their portfolios may be more important.

Of course, while different decision-makers have different needs for impact information (i.e. timeliness, accuracy, etc.), large errors in impact estimates can have significant negative influences on specific decisions. Conversely, certain decisions can be improved with accurate impact information. Two scholars have asked "how do we determine whether a model has 'correctly' simulated an impact?" (West and Lenze 1994). They find that "at present, most evaluation in regional impact analysis is confined to the fairly simple and non-rigorous step of asking whether the results look 'reasonable'" and they recommend further research in the area of model evaluation.

Add in climate variability, change, and forecasts

The three risk assessment methods reviewed above will work best if the documented past is a reliable guide to the near future. But what if the past is not a reliable guide to the future? It has recently been appreciated that the climatological dimensions of tropical cyclone impacts may undergo significant variability on annual and decadal time-scales (and perhaps even longer scales related to climate changes; Diaz and Pulwarty 1997). Further, in recent years, scientists have begun to demonstrate some skill in interannual climate forecasts (e.g. Landsea et al. 1994; O'Brien, Richards and Davis 1996; Glantz 1996).

Improved understanding of climate variability and ability to anticipate changes in event frequency, magnitude, and location have two important implications for efforts to better understand tropical cyclone impacts. First, variability on annual and decadal time-scales means that the 120 or so years of reasonably good data on the climatology of hurricanes may not be a reliable guide to the immediate future. Second, the presence of forecasts of tropical cyclone activity provides risks and opportunities for decision-makers. They present opportunities for decision-makers to alter their decision routines in such a fashion as to incorporate new information into their plans and policies. The forecasts also present risks because they are largely untested and the costs of an improperly used prediction can be very large.

There are two additional methods of risk assessment that seek to utilize information beyond that available in the historical record (Dlugolecki et al. 1996).

- *Event prediction*
  With reliable information of future events, decision-makers can implement policies to take advantage of that knowledge. There is a large literature on the use of climatological information by decision-makers in, for example, agriculture and utility industries (cf. Changnon, Changnon and Changnon 1995), but very little (if any) in the area of tropical cyclones. The use of climate forecasts of tropical cyclones by decision-makers requires research attention.

• *Parallelisms*
In cases where data on or experience with extreme events are lacking in a particular region, knowledge of a similar place or time might substitute. As Glantz (1988) has suggested: "in order to know how well society might prepare itself for a future change in climate (the characteristics of which we do not yet know), we must identify how well society today can cope with climate variability and environmental impacts". This approach is also called "forecasting by analogy". An example is Coch and Wolff (1990), who explore the "probable effects of a storm like Hurricane Hugo on Long Island, New York".

Recent experience with tropical cyclones illustrates the co-dependency of physical and societal dimensions of tropical cyclone impacts. Over the past several decades of relatively depressed hurricane activity, property losses have risen exponentially with the period 1990–1995 seeing more damages (inflation adjusted) than the 1970s and 1980s combined. The increase in property losses has been driven by increased development in exposed coastal locations. However, as noted in the introduction to this chapter, some decision-makers have incorrectly interpreted the trend of increased damages to indicate more storms. Others have interpreted a trend of fewer casualties related to hurricanes as an indication that hurricanes no longer pose a serious threat to life.

The main lesson of these errors of interpretation is that decision-makers need to pay explicit attention to the interrelation of physical and societal aspects of hurricanes. It is important to be able to look at the "big picture" of tropical cyclone impacts in order to avoid analytical blind spots and misinterpretations of selected trend records.

Conventional risk assessment methodologies have important intrinsic limitations and impact models have potential for significant blind spots. Further, West and Lenze (1994) argue that:

> ... such research must also include non-modeling impact methodologies. Indeed, the uncertainties and complexities of the issue may ultimately yield the outcome that impact techniques derived from adapting historical experience with other natural disasters or using quasi-experimental control group methods yield better projected impacts.

To understand the "big picture" of impacts, decision-makers need a framework within which they can balance and assess the relative physical and societal aspects of tropical cyclones. With a broad context in view the decision-maker will be less likely to succumb to analytical blind spots. In the past, for the most part, the broad context of tropical cyclone impacts has been defined by trend data on hurricane events, property losses, and casualties.

Obviously, such trend data are not a reliable indicator of the causes of impacts. What is needed is a conceptual framework within which tropical cyclone impacts and the various methodologies used to interpret them can be broadly understood. The concept of vulnerability provides such a framework.

2.3.2 Societal vulnerability to tropical cyclones: a framework for assessment

Clearly, there are a wide range of factors important to understanding tropical cyclone impacts. Further, different decision-makers will need to understand impacts in different ways, depending upon the particular policy decisions that they face. This section uses the concept of "vulnerability", as described in Section 2.2, as an integrative framework within which one can more reliably assess preparedness for tropical cyclone impacts in broad context.

The impacts of tropical cyclones on society are a result of both physical and societal factors. If the term *event incidence* is used to refer to the physical aspects of tropical cyclones and *societal exposure* is used to refer to social conditions, then *vulnerability* to tropical cyclones can be defined as a function of incidence and exposure. Thus, vulnerability of society to hurricanes can be defined as follows:

$$\text{vulnerability} = f(\text{incidence, exposure})$$

Society mitigates its exposure, and consequently impacts, through conscious action which we call "preparedness". Preparedness reduces exposure, and consequently vulnerability. The concept of vulnerability provides a framework for an assessment of tropical cyclones and can suggest causes of impacts, consequences, and alternative actions for improving preparedness.

Vulnerability assessment requires knowledge of incidence, exposure, and preparedness. Incidence is determined through climatology and meteorology. Factors that are important in assessing incidence include frequency, magnitude, location, intensity, etc. Exposure is determined through assessment of people and property at risk to the event. Assessment of both tropical cyclone incidence and exposure to tropical cyclone impacts can be both backward-looking (e.g. the climate record, demographic trends) and forward-looking (i.e. forecasts, projections). Preparedness assessment is more difficult to perform and is consequently least often undertaken. It involves asking questions such as "How prepared are we?" and "What steps can be taken now to reduce the impacts of future events?" The answers to these questions are of course intimately related to knowledge of incidence and exposure.

Vulnerability is a useful concept because it helps the analyst to recognize explicitly that there are both physical and social causes of impacts. It is more difficult to discern the balance of physical and social factors driving impacts from simple trend data on losses or casualties. For instance, some, like the US

Congress, have concluded that the reason for the observed rise in damages must be more frequent and powerful storms. Yet what has changed over the past several decades is the amount of people and property at risk to hurricane impacts in coastal locations. It is also arguable that during this period resources devoted by society to hurricane mitigation efforts have not kept pace with development.

Vulnerability assessment is by definition a multidisciplinary process. A reliable assessment of a society's vulnerability to hurricanes would necessarily include the following disciplines or representatives of areas of knowledge: climatology, meteorology, wind engineering, psychology, sociology, political science, emergency management, insurance, elected and appointed officials, the general public, transportation, responsibilities and capabilities of various federal, state, and local government agencies as well as private organizations, etc. More could surely be added to this list. Vulnerability assessment is also an integrative process – it brings together disparate knowledge in a manner that allows for identification of problems and insights as to their resolution. With the inherent difficulties involved with multidisciplinary, integrative efforts, it should not be a surprise that this sort of assessment has rarely been completed in the past.

2.3.3 Incidence assessment

In terms of incidence, climate history provides a first approximation to hurricane occurrence, intensity, and landfall frequency (probability) for a particular community. On a year-by-year basis, work on seasonal variability has the potential to improve upon data gleaned from the historical record. For decisions that cover more than one hurricane season, e.g. building a house, the historical record remains the best source of information on hurricane incidence. Climate statistics and experience provide decision-makers with a range of information (e.g. annual probabilities of landfall) that can be used to structure action alternatives (e.g. building codes). Knowledge of hurricane incidence is a central factor in the determination of vulnerability of people and property to hurricanes.

2.3.4 Exposure assessment

Determining population and property at risk

With knowledge of hurricane incidence, it is possible to determine people and property at risk to hurricanes (as described in Section 2.2.3). In defining "risk" it is important to recognize that the "choice of definition is a political one, expressing someone's views regarding the importance of different adverse effects in a particular situation" (Fischhoff, Weston and Hope 1984). Population and property at risk can be defined in a number of ways. For instance, one measure of risk is landfall frequency based upon the climatological record. Risk can also be defined analytically, e.g. through scenarios that move storms to different times and locations. For instance, what would the effects of the 1938 New England hurricane be in 1998? Risks can also be evaluated in a more subjective manner, such as occurs when an individual decides whether or not to evacuate.

In practice, determination of population and property at risk at particular locales involves the assessment of coastal inundation due to storm surge based upon computer models of landfalling storms. People vulnerable to storm surge are generally considered to be most at risk due to a hurricane landfall. Hurricane winds and flooding also threaten people and property. Based upon determination of people and property at risk (however defined) actions are taken to reduce the risk. For example, building codes are designed based upon a community's expectation of hurricane incidence, and evacuation plans are also designed based upon climatological factors. Under the conventional definition of the hurricane problem, hurricane policy is centered on reducing "risk" associated with the storm's impact. In general, however, such policies lack a means to fully integrate existing preparedness efforts in the definition of hurricane risk. One example of an effort to include preparedness in definition of risk is the Building Code Effectiveness Grading Schedule (BCEGS) created by the Insurance Institute for Property Loss Reduction (BTFFDR 1995). The BCEGS is a ranking of community implementation of building code *enforcement* based on their budget allocated for that purpose and the qualification of their inspectors.

Preparedness assessment

In order to assess a community's preparedness, it is necessary to assess the "health" of the various decision processes in place to reduce vulnerability. A healthy decision process will meet criteria of content and procedure (see Lasswell (1971) for a working list of such criteria). Assessment of content is relatively straightforward: Is there an up-to-date SLOSH model analysis? Is there an evacuation plan? Are there shelters? Do building codes accurately consider climatology? Assessment of the procedural aspects of decision processes is a more daunting task, requiring close attention to the who, what, where, when, and why of particular decisions.

In order to assess the "health" of a particular decision process, one must first have a "map" of the process with respect to important decisions relevant to reducing vulnerability. In the context of an approaching hurricane, important decisions might include whether or not to order an evacuation (and who to evacuate, when), how to enforce building codes, or the amount of resources to put into education of the public about the hurricane threat. In 1992, Lee County Florida prepared a decision process map which can be used as a starting point in efforts to determine the health of preparedness (see

Appendix E). Such a map, customized for the decision routines of a particular user, provides a useful starting point for the analyst seeking to answer the question "How prepared are we?" Of course, assessment of an entire community's preparedness is a significant task that would require attention and resources.

CHAPTER 3

Tropical Cyclones on Planet Earth

3.1 LIFE OF A HURRICANE

Tropical cyclones are relatively rare weather events, with only about 84 per year over the entire Earth. During the period 1968–1989, the Atlantic Ocean basin averaged 9.7 tropical cyclones annually, of which an average of 5.4 became hurricanes (Neumann 1993).

3.1.1 Birth and growth

Hurricanes typically form over warm oceans from pre-existing regions of relatively low surface pressures with an associated cluster of thunderstorms (Riehl 1954). Regions of relatively low and high pressure exist in the tropics due to differences in the weight of the atmosphere over these different locations. Such pressure systems are analogous to those shown as "L" and "H" on television weather reports. These regions are formed by mechanisms that include differences in the heating of the atmosphere over the land and ocean surface, and the flow of air over surface features such as mountains. Another, more complicated factor in the development of areas of relatively high and low pressure is associated with propagating atmospheric waves that develop due to differences in the temperature of the atmosphere and/or wind across a region.

Air tends to move towards the center of the low pressure region (and away from high pressure). However, in the Northern Hemisphere, the rotation of the Earth deflects the air such that it tends to move in a counterclockwise direction around a low pressure system (and clockwise around a high pressure system; Figure 3.1) as it spirals toward the center of the system. This wind deflection results from the Coriolis effect, as schematically shown in Figure 3.2. As air spirals over the warm oceans toward the center of the low pressure, the ocean provides water vapor (through evaporation) and the warm water surface adds heat to the lower atmosphere (Ooyama 1969; Malkus and Riehl 1960). This movement of water vapor into the thunderstorms within the low pressure area from the outside environment is essential to tropical cyclone

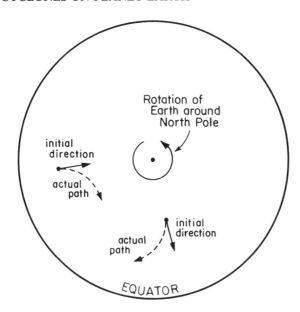

Figure 3.1 Deflection of wind by the Coriolis effect. Wind flows toward a low pressure, and outward from high pressure and is deflected to the right by the Coriolis effect. This is why air tends to circulate clockwise around high pressure systems, and counterclockwise around low pressure systems in the Northern Hemisphere

formation, as precipitation substantially exceeds local evaporation from the ocean (Liu, Curry and Clayson 1995).This evaporation and heating provides energy for thunderstorms.

As air spirals inward, it is forced upwards since the air cannot go down into the ocean, nor compress significantly. This process is called low-level wind convergence, and it results in a more favorable heat and moisture environment for thunderstorms. A visible result is the further development of cumulus clouds. If the middle and upper troposphere are sufficiently moist, these cumulus clouds can grow to the top of the troposphere (Figure 3.3). These thunderstorms effectively move heat and moisture to the upper levels of the troposphere, which allows for further development.

Since the layer of atmosphere above the troposphere, i.e. the stratosphere, acts as a barrier to the heat and moisture that is moving upwards through the atmosphere, the air that is transported upwards spreads out horizontally (called upper-level wind divergence), which serves to lower pressures further near the surface. Decreased pressure near the surface results in additional low-level wind convergence, thereby enhancing conditions for additional thunderstorm development. This additional pressure drop occurs due to less air converging into the center near the surface relative to the divergence of air in the upper troposphere. Low vertical wind shear permits the thunderstorms to

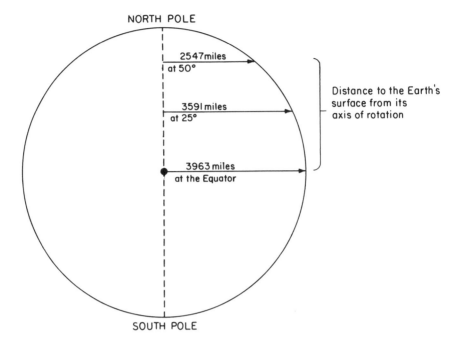

Figure 3.2   Illustration of the axis of the Earth's rotation and distance to the Earth's surface. Note that the change of distance between the Earth's axis of rotation and the surface for a fixed change of latitude increases as you move poleward. This is why the Coriolis effect becomes larger towards the poles

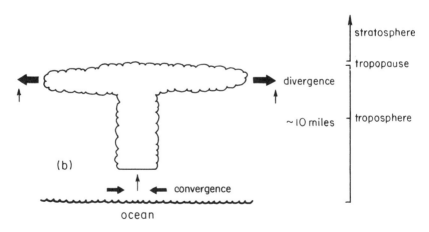

Figure 3.3   A simplified illustration of the development of organized thunderstorm activity and associated lower tropospheric wind convergence, and upper tropospheric wind divergence. When more air is removed aloft than is replaced at low levels, the surface pressure falls at the center of the developing tropical cyclone

remain concentrated. When large vertical wind shear is present, the thunderstorms associated with the surface circulation are blown downstream, leaving the low-level center isolated from the thunderstorms (Aberson and DeMaria 1994). Figure 3.12 illustrates how vertical shear influences the development of thunderstorms associated with a tropical low pressure system.

This process of low-level wind convergence and thunderstorm development in the middle and upper troposphere, plus upper-level wind divergence, has been described as a "thermal heat engine". A pioneering study of this coupling between the lower and upper troposphere is given in Riehl (1948). A thermal heat engine can be described as a fire in a fireplace, in which the warm air from the burning wood rises up a chimney, with low-level fresh air feeding the fire. In a developing tropical system, thunderstorm formation is analogous to the wood in the fire. The thunderstorm produces heat in the process of converting water vapor to liquid water and ice as the air rises. Technically the conversions (and release of heat) from these processes are called condensation (for water vapor to liquid) and deposition (for water vapor to ice). The heat from this phase change is referred to as latent heating. Pressure will continue to fall at the surface in the center as long as the divergence aloft is greater than the convergence at low levels. Generally, the air which diverges aloft sinks at the periphery of the cluster of thunderstorms, with some of the air recycled back into the storm. If a significant quantity of the air is recycled, then the pressure cannot drop very far (Merrill 1985; Petrosyants and Semenov 1995). In addition, the subsiding air warms and dries the region surrounding the cloud system, providing an atmosphere which is less conducive to thunderstorms. This is why developing tropical systems are usually surrounded by a clear, cloudless area.

At this point the developing system appears as an unorganized cluster of thunderstorms with only a weak surface wind circulation. It is called a tropical disturbance. If conditions for development remain favorable, including low vertical wind shear, warm ocean surface, and sufficient moisture throughout the troposphere, they may further develop into a tropical depression when the sustained surface winds, circulating around the center, strengthen up to no more than 39 miles per hour (18 meters per second) (see Emanuel (1988) for an overview of this process of development). At this point, hurricane forecasters number the depression in order to identify it for further observation. The winds in a tropical system result from air seeking to flow from regions of relatively high pressure to regions of relatively low pressure. Because of the circulation of the system, due to the rotation of the Earth, the winds spiral in, giving the developing storm its characteristic spiral appearance.

Meanwhile, as the storm is developing, it continues to move over the open ocean. In the Atlantic Ocean Basin, a typical track for the developing storm is east to west from off the coast of Africa toward the West Indies and North American continent.

Further development, such that sustained winds reach at least 39 miles per hour (18 meters per second), and no more than 74 miles per hour (33 meters per second), results in the system being classified as a tropical storm. At this point, forecasters name the tropical storm. At tropical storm strength, winds become a substantial threat to coastal locations; however, winds and precipitation in tropical depressions can also threaten marine interests, and inland and coastal communities.

In the Atlantic, on average, only about five of around 20 tropical depressions reach hurricane strength (less than half of tropical depressions reach tropical storm strength).

3.1.2 Maturity

As the low-level flow becomes stronger and a tropical storm strengthens, it becomes increasingly difficult for air to reach all of the way into the center. This limitation occurs because the winds become stronger as they spiral into the center, resulting in a countervailing tendency for the air to be spun out of the inward spiral. This is called centrifugal force.

The spiraling in of airflow at low levels results in the formation of the tropical cyclone's characteristic thunderstorm "spiral bands". The inflow becomes enhanced in several spiral bands because thunderstorms, as they are generated, tend to enhance convergence in their vicinity, resulting in a more favorable environment of moisture and temperature for subsequent thunderstorm development than that found in the inflow region without clouds. This positive feedback between enhanced local convergence and thunderstorms acts to perpetuate the spiral bands. The spiral bands tend to rotate slowly counterclockwise (Northern Hemisphere), at less than the speed of the wind, around the center of the storm. These spiral bands are also called "feeder bands" because of their contribution of heat and moisture to the center of the storm. Figure 3.4 illustrates from satellite imagery the form of these spiral bands as seen from space.

When minimal hurricane force winds are achieved, air can no longer reach the center of the low pressure, resulting in a more or less calm area called the eye (Malkus 1958). The presence of an eye may be one reason for the definition of a hurricane as having winds 74 miles per hour (33 meters per second) or greater. A tropical storm has winds less than 74 miles per hour and typically does not have an eye. The region called the eye wall surrounding the eye corresponds to the maximum inward penetration of the inward spiraling air and is the region of strongest winds. The development of an eye is schematically illustrated in Figure 3.5. Figure 3.6 illustrates the appearance of an eye as viewed from a geostationary satellite. Note the well-defined hole in the center of the cloud mass which is associated with the hurricane. A computer simulation of flow into, upward, and out from the eye wall is shown in Figure 3.7.

Figure 3.4   An infrared image taken by a geostationary satellite of Hurricane Edouard on 26 August 1996. Photo courtesy of Ray Zehr, NOAA

Inside the eye, air sinks in a manner analogous to the subsidence that leads to the cloud-free area around the storm, although here it is concentrated in a small area. This sinking warms and dries the air. Since warmer air can contain larger amounts of water vapor before condensation must occur, clouds tend to dissipate in the eye. A result is the characteristic cloud-free eye. In intense and intensifying hurricanes, the eye region can be completely cloud-free.

In very intense hurricanes such as Gilbert (1988), double, concentric eye walls can form when an inner eye wall weakens as the inflow to the storm, unable to penetrate as close to the center because of the strong winds, begins to be concentrated in an outer eye wall. The strongest winds in the hurricane decrease as a result, because the difference in pressure between the eye and the outer eye wall is less. If the outer eye wall subsequently contracts in radius, the hurricane winds can again intensify until the inflow can no longer reach the tightly wound eye wall, and a new outer eye wall develops. This cycle of creation and weakening of eye walls results in fluctuations in the maximum winds in the storm. Willoughby, Clos and Shoreibah (1982) and Willoughby (1990) provide more details on concentric eye wall cycles.

Individuals who experience the passage of the eye are often surprised by the appearance of blue, sunny skies (or stars at night) along with a dramatic cessation of the wind within the eye. Unfortunately, this interlude is usually

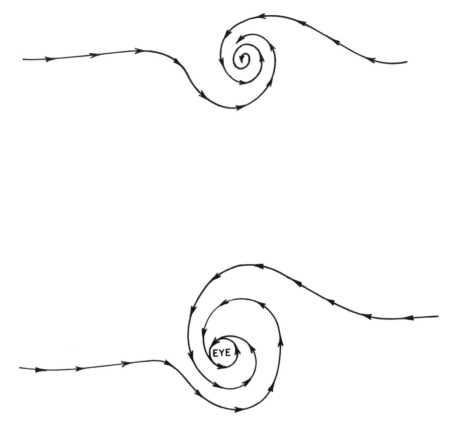

Figure 3.5   Schematic of the formation of an eye, as the winds become too strong to permit the air to spiral all the way into the center of the tropical cyclone

followed by a rapid increase of the wind from the direction opposite to that observed before the passage of the eye.

Figure 3.8 illustrates the change of wind speed, temperature, and pressure from the periphery of a hurricane into its center (at a height of about 3 km above the ocean). This cross-section, obtained by averaging 16 aircraft penetrations into Hurricane Anita in the Gulf of Mexico on 2 September 1977, is typical of a strong, mature hurricane. Note that the radius of maximum winds of about 144 miles per hour, which occurred about 13 miles (20 km) out from the center, corresponds to the outer limit of the warming of up to 9°F (5°C) caused by the sinking of air in the eye. The average upward motion of air in the eye wall is also largest at the radius of maximum winds.

Figure 3.6   Visible satellite image of the hurricane eye in Hurricane Edouard on 26 August 1996 (see Figure 3.4). At this time, sustained surface winds were about 140 mph (224 kph). Photo courtesy of Ray Zehr, NOAA

Figure 3.9 illustrates a computer model simulation of a vertical cross-section through a hurricane. Shown in this figure, and observed in real storms, is the typical increase in diameter of the eye with height due to the tilt of thunderstorms in the upper troposphere outward from the center of circulation. One reason that this tilt occurs is the strong outflow from the top of the storm. Figure 3.10 presents a horizontal cross-section of the simulated hurricane with the eye wall and spiral bands more clearly defined.

The view from within the eye can be spectacular in strong hurricanes. The deep thunderstorm clouds of the eye wall have been characterized as appearing like a gigantic rotating coliseum. Birds have been reported as finding sanctuary within the storms, often being transported hundreds or even thousands of miles from their native regions. Ships within the eye often report numerous birds perching on their vessels in order to rest.

Robert H. Simpson graphically describes the appearance of the eye as seen from an aircraft penetrating into Typhoon Marge:

Soon the edge of the rainless eye became visible on the (radar) screen. The plane flew through bursts of torrential rain and several turbulent bumps. Then suddenly we were in dazzling sunlight and bright blue sky.

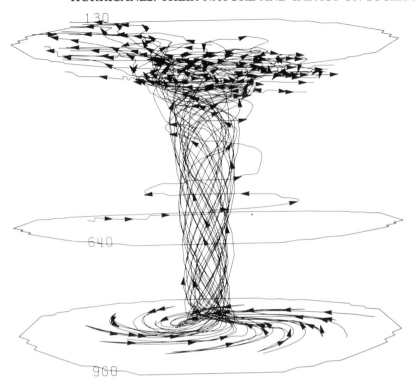

Figure 3.7   A model simulation of the behavior of winds in a hurricane. Note the low-level convergence and upper-level divergence. The numbers 960, 640, and 130 on the figure refer to the atmospheric pressure in millibars at the different levels in the storm. Source: Anthes and Trout (1971). Reprinted with permission from Weatherwise

Around us was an awesome display. Marge's eye was a clear space 40 miles in diameter surrounded by a coliseum of clouds whose walls on one side rose vertically and on the other were banked like galleries in a great opera house. The upper rim, about 35,000 feet high, was rounded off smoothly against a background of blue sky. Below us was a floor of smooth clouds rising to a dome 8000 feet above sea level in the center. There were breaks in it which gave us glimpses of the surface of the ocean. In the vortex around the eye the sea was a scene of unimaginably violent, churning water. (Simpson 1954)

Figure 3.11 illustrates an average relation between wind speed in an Atlantic hurricane and its central pressure. Atmospheric pressure is typically measured in units of millibars; where a thousand millibars corresponds to about the average pressure at sea level. A very intense hurricane will have a central pressure of about 900 millibars, or about 10% less than a typical sea level pressure (see Table 1.1). Individual hurricanes, however, will deviate

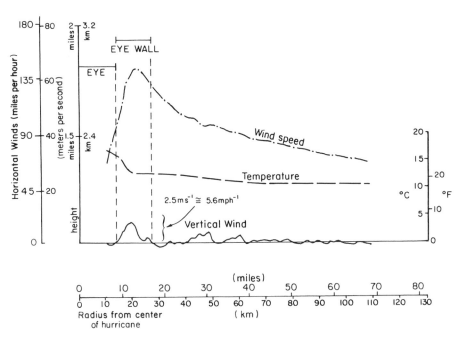

Figure 3.8   The structure as a function of distance from the storm center as approxi-
mated by averaging 16 profiles for Hurricane Anita on 2 September 1977. The
temperature line illustrates the warm core of the hurricane. The eye wall is the zone
with the greatest sustained wind speeds and the strongest averaged upward motion.
Source: Willoughby (1979)

from the relationship illustrated in Figure 3.11. For instance, in two storms
with identical central pressure, a larger storm will tend to have weaker winds
than a small storm because the horizontal difference in pressure (which drives
the winds) in the larger storm is generally spread over a greater distance.

The deepest pressure, and hence maximum wind, that is possible in a
hurricane is well correlated with ocean surface temperature (Miller 1958).
Using the analogy of a heat engine, wind circulation would be expected to be
stronger when the heating from the ocean is greater. A hurricane, for example,
would not, in general, be expected where the ocean surface temperature was
less than 79°F (26°C) (K. Emanuel of the Massachusetts Institute of Technology produces graphs of maximum expected hurricane intensity based on sea
surface temperature, which are available on the www at http://grads.iges.org/
pix/hurrpot.html). One of the lowest pressures ever observed in an Atlantic
hurricane (892 mb) was in the Labor Day, Florida Keys Storm of 1935 in
which 408 deaths occurred. If Hurricane Andrew (1992) had more time to
intensify, then it could have had a significantly lower central pressure at

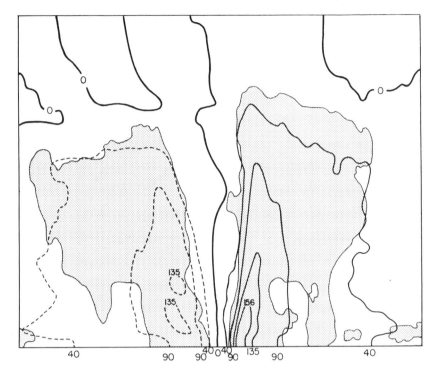

Figure 3.9   A model simulation of the eye and eye wall region of a mature hurricane. The contours show wind speed in miles per hour. The solid contours represent winds moving into the page and dashed contours represent winds moving out of the page. The shaded region illustrates the locations of cloud. Figure provided by Dr Mel Nicholls, Colorado State University; this work is also reported in Nicholls and Pielke (1995)

landfall because of the warm water of the Gulf Stream (Willoughby and Black 1996).*

Fortunately, few hurricanes attain their theoretical maximum potential intensity. The presence of significant vertical wind shear is likely the major reason that hurricanes do not attain this level of intensity.

---

* There are several theories on the maximum potential intensity of tropical cyclones (Gray 1997). The heat engine analog of these storms has been used to estimate maximum possible strength based on inflow and outflow temperature differences (Emanuel 1986; 1987). Emanuel et al. (1995) use this concept to propose that extremely strong hurricanes could develop, which they call "hypercanes". A different mechanism has also been proposed where hurricane intensity is limited by controls on deep cumulus cloud buoyancy in the eye wall (Holland 1997). William Gray of Colorado State University, however, maintains that upper-level momentum export from the storm, along with surface frictional dissipation must be considered together with the thermo-dynamic controls proposed by Emanuel and by Holland (Gray 1997).

240 miles

Figure 3.10   Horizontal cross-section through a simulated hurricane at a height of ~3 miles (5 km). The contours clearly outline the eye wall with the most intense ascending air on its south side, as well as in the spiral bands feeding into the storm. Figure provided by Dr Mel Nicholls, Colorado State University; this work is also reported in Nicholls and Pielke (1995)

3.1.3 Decay

A mature hurricane can cover thousands of square miles and is a striking atmospheric feature as viewed from space. On the Earth's surface, the mature hurricane can lead to great property damage and loss of life. Fortunately, mature hurricanes last for only a limited time as decay is inevitable. Decay can occur either over the ocean or over the land.

Over warm ocean waters, an increase in the vertical wind shear because of the movement of the storm into a high shear environment can disrupt the hurricane by reducing the effectiveness of the linkage performed by thunderstorms between the lower and upper troposphere (see Figure 3.12). In some cases, strong wind shear can even rip a hurricane apart. This is a primary reason for hurricane decay over a warm ocean. Three other factors can contribute to decay. The first is the movement of dry air into the center of the storm, therefore inhibiting thunderstorm development. This could occur due to the movement of a dry air mass off a continent into the spiral inflow of the hurricane. A second reason is the passage of the hurricane over a mountainous island such as Cuba or Hispañola, which can disrupt the inflow into the center

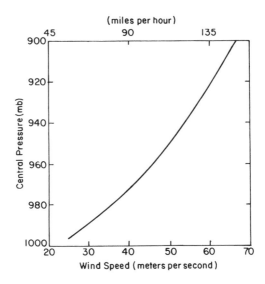

Figure 3.11   The average relation between wind speed and central pressure in an Atlantic hurricane. Source: Simpson and Riehl (1981). Reproduced by permission of the Louisiana State Press

Effects of Vertical Wind Shear on Tropical Cyclones

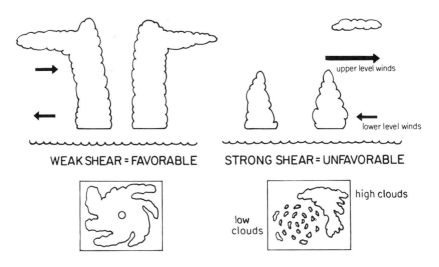

Figure 3.12   A schematic illustrating the effect of vertical shear of the horizontal wind on tropical cyclone development. Figure provided by Stanley Goldenberg and Neal Dorst of the Hurricane Research Division AOML/NOAA

of the storm. A disruption in the outflow from the hurricane, due to the upper-level divergence being replaced by upper-level convergence, as might occur due to large-scale atmospheric circulation changes, can also weaken and even dissipate a storm.

Hurricanes also weaken rapidly when they traverse over cool water or inland after they make landfall (Kaplan and DeMaria 1995). There are two major reasons for the decay of hurricanes over a cool ocean or after landfall. First, a hurricane requires that the warmest temperatures which are associated with the storm be in its center in order for the heat engine to work. However, as air spirals into a hurricane, it expands as a result of the lower pressures closer to the eye. Unless heat is added, this expansion results in cooling. (The same process occurs when air is let out of a tire. The expanding air at the nozzle from the pressurized tire is substantially colder than the surrounding air.) The cooling works against maintaining the heat engine.* This cooler air is not a favorable environment for continued thunderstorm development. Over land and cool ocean surface waters, there is no heat source to counteract this cooling. The direct result is that thunderstorms weaken within the eye wall. The low central pressure of the tropical cyclone correspondingly rises as the coupling between the lower and upper troposphere is reduced and, as a result, the divergent winds in the upper levels of the tropical cyclone are diminished in strength. Thus, the eye wall tends to be destroyed as the hurricane weakens to tropical storm strength. The second reason that hurricanes decay over land and cold oceans is that they lose the unlimited source of water vapor from the warm ocean water which is essential for maintaining the strong thunderstorms in the eye wall region.**

Wind speeds also decrease near the ground after landfall. While not directly related to a hurricane's decay, this decrease in wind speed near the ground is due to the aerodynamic roughness of the land. Trees, buildings, and even grasslands tend to be rough surfaces compared to the ocean, with the result that wind is decelerated. One major result of this difference in roughness is that, even if the wind well above the surface remained constant at landfall, the greater retardation of the flow by the rougher surface over land would result in slower wind speeds at the surface.

The cooler air over land magnifies this reduction in wind speed near the surface even further. Therefore, while the winds well above the ground may even accelerate, as those levels tend to become decoupled from frictional retardation by the surface, the winds near the surface can become quite weak. While this decoupling of near surface flow from the winds aloft does not

---

* For example, air which originates at a pressure of 1000 mb and 80°F (27°C) in the region surrounding the storm would cool to 64°F (18°C) at a pressure near the center of a storm of 900 mb, unless heat were added.

** A surface of water evaporates at a rate which is directly related to its surface temperature. For the same amount of moisture just above the surface, and a surface pressure of 1000 mb, an ocean area with a temperature of 80°F (27°C), for example, will evaporate at a rate about 64% greater than when the surface is at 64°F (18°C).

directly reduce the overall intensity of a hurricane, its destructive potential near the ground is reduced.

3.1.4 Criteria for development and intensification of a tropical cyclone

There are several major criteria for the development of tropical storms and hurricanes:

1. The presence of a pre-existing low surface pressure.
2. A warm, moist tropical atmosphere that is conducive for thunderstorm development.
3. Ocean surface temperatures greater than about 79°F (26°C) so that sufficient moisture and heat can be supplied into the low pressure area in order to sustain the thunderstorm development (Palmén 1948).
4. Weak vertical wind shear (less than about 17 miles (27 kilometers) per hour) between the upper and lower troposphere such that the developing thunderstorms remain over the region of lowest pressure.
5. A distance sufficiently removed from the Equator (generally by more than 4°–5° of latitude) so that air will tend to spiral inward at low levels towards the lower pressure, and outward at upper levels away from high pressure. Near the Equator, there is little rotation associated with low and high pressure systems. At and near the Equator, areas of thunderstorms occur without substantial pressure falls and without strong, sustained, low-level convergent winds. In this region, the converging winds are able to eliminate the low pressure system after a short period of time through the movement of air into the low pressure center. Such transient thunderstorms occur frequently within about 4° or 5° of the Equator.
6. The development or superposition of high pressure in the upper troposphere over the surface low so as to evacuate air far from the region of the cyclone, thereby permitting surface pressures to continue to fall. This criterion is particularly important in the transition of a tropical storm to a hurricane, and to further intensification.

Once a tropical cyclone reaches the intensity of a hurricane, it will not weaken unless:

• The vertical wind shear becomes too large
• Its source of heat and moisture is reduced as a result of passage over land or relatively cold water
• Dry, cool air, which does not favor the development of thunderstorms, is transported into the system
• The high pressure aloft is replaced by a cyclonic circulation which adds air to, rather than evacuates air from, the hurricane heat engine. Since thunderstorms tend to perpetuate high pressure aloft, larger-scale atmospheric circulation changes are required to remove such an outflow region.

Note that the requirements for the continuance of a hurricane are less restrictive than those for its development. Therefore, while only occasionally are atmospheric conditions ripe for the genesis and development of a hurricane, once established it tends to be a persistent weather feature.

In contrast to the locations for tropical cyclone development, the ability of a hurricane to persist once developed, as long as the criteria listed above do not occur, accounts for the spread of storm tracks well beyond their source region, as evident in Figure 3.13.

3.2 SPECIAL CASES OF DEVELOPMENT AND INTENSIFICATION

Tropical cyclones can also develop from mid-latitude, low pressure systems outside the tropics. These include low pressure centers that become completely cut off from the belt of upper tropospheric winds that are referred to as the jet stream (Sadler 1976; Pfeffer and Challa 1992; Ramage 1959; Montgomery and Farrell 1993). In contrast to tropical low pressure systems, these mid-latitude cut-off lows initially have colder air near their center. For this reason, these lows are referred to as being "cold-core lows". These atmospheric regions of disturbed weather, referred to as "subtropical cyclones", can transition into a tropical cyclone as a result of thunderstorms in their center.* In a tropical cyclone, the warmest air is near the center. This structure of a warm core, coincident with the low pressure center, permits the weather system to work as a direct thermal heat engine. If the thunderstorms persist long enough over the center of lowest surface pressure, a tropical cyclone can develop from the cold-core low.

Occasionally, as with Hurricanes Ivan and Karl in 1980, the tropical cyclones can develop even over water of less than 79°F (26°C; Lawrence and Pelissier 1981). One explanation is that at higher latitudes and with larger and more intense cyclones, the vertical shear of the horizontal wind has less of an effect on intensity change than at lower latitudes and for small weak storms (DeMaria 1996). Therefore, even at temperatures colder than 79°F (26°C) at higher latitudes, large initial circulations with relatively weak clusters of thunderstorms can still spin up into a tropical cyclone.

As an alternative explanation, tropical cyclone genesis and intensification can occur irrespective of the ocean surface temperature whenever the atmosphere is conducive to thunderstorms, and the vertical wind shear is small. Generally, over oceans colder than 79°F (26°C), these two conditions are not simultaneously met (C. Landsea, personal communication).

---

* Since 1968 the tracks of subtropical cyclones were included in the annual summary of tropical cyclone activity. These are referred to as subtropical depressions (winds less than 34 knots) and subtropical storms (winds 34 knots and greater) (Neumann 1993).

84

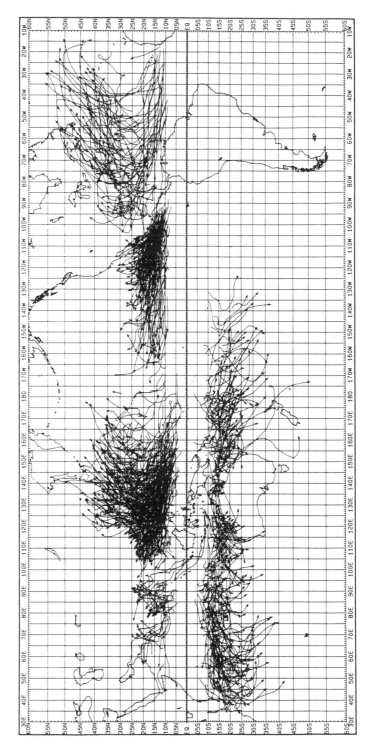

Figure 3.13 Global tropical cyclone tracks (1979–1988). Source: Neumann (1993)

In the polar regions, there have even been low pressure systems documented that have many of the characteristics of tropical lows (Forsythe and Vonder Haar 1996; Rasmussen 1985). Called "polar lows", these features have been found to have warm cores and thunderstorms despite being over ocean waters that are well less than 79°F (26°C). The deep cumulus clouds are able to form even in this cold environment because of the presence of very cold air above the ocean surface. The large difference in temperature between the ocean and the air provides the evaporation and heat needed to a sustain the thunderstorms.

3.3 GEOGRAPHIC AND SEASONAL DISTRIBUTION

3.3.1 Origin

Tropical cyclones have been given different names depending on their region of origin. In the western north Pacific, they are called typhoons, while in the Bay of Bengal (east of India), they are referred to as severe cyclonic storms of hurricane intensity. In the Atlantic, Gulf of Mexico and Caribbean, they are hurricanes. Evidence of tropical cyclones have been documented in a variety of other geographic locations including Europe and north Africa at earlier geologic times (Ager 1993).

Typically, in the Atlantic Ocean Basin, tropical storms and hurricanes develop over warm water between around 10°N and 35°N, generally during the summer and fall (Figure 3.14a and b). During an average year about 16 tropical cyclones develop in the eastern Pacific and approximately 10 in the Atlantic including the Gulf of Mexico and Caribbean Sea (Neumann 1993). During the period of record, tropical cyclones have never developed south of the Equator in the western hemisphere east of 130°W because of one or more of the following factors: the relatively cold ocean temperature, typically strong winds in the upper troposphere, or the absence of an initiation area for tropical low pressure systems with an associated cluster of thunderstorms (Gray 1968). (McAdie and Rappaport (1991), however, discussed the formation of a weak tropical cyclone in the south Atlantic west of tropical Africa in 1991.) Elsewhere these storms develop in the Indian Ocean, western Pacific, and eastern Pacific north of the Equator (see Figure 3.13 and Appendix D). The western north Pacific is the most active area, with an annual average of more than 26 tropical cyclones. Chan and Shi (1996) have found that these storms have occurred more frequently since the late 1980s. Globally, there are about 84 tropical cyclones with an annual average of 45 that reach hurricane strength (Neumann 1993).

3.3.2 Movement

For human societies, a tropical cyclone's track, or motion, is more important than the location of origin. Figure 3.13 illustrates the path of all tropical

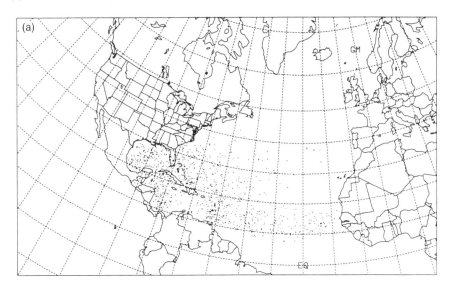

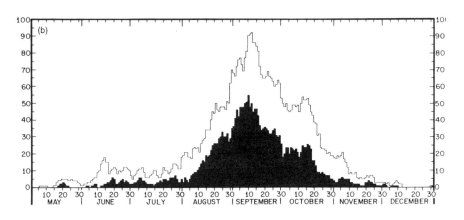

Figure 3.14 (a) Locations where tropical storm intensity was attained in observed storms for the period 1886–1995 (prepared by J. Eastman). (b) Seasonal distribution of tropical storms and hurricanes in the Atlantic Ocean Basin (World Meteorological Organization 1993)

cyclones globally for the period 1979–1988. While it is difficult to track individual storms from this figure, regions of high frequency of occurrences of storm passage are easily viewed by their large concentrations of tracks. Correspondingly, areas of infrequent but occasional storm passage are clearly shown.

The area of damaging winds extends well beyond the point location of hurricane position that is displayed in track maps such as shown in Figure

3.13, although the region of most destruction potential is concentrated close to the center of the storm in the region of the eye wall. The largest radius of tropical storm winds (greater than 39 mph) was 680 miles (1100 kilometers) in Typhoon Tip in the western Pacific in October 1979 (Holland 1993a). The smallest was Tropical Cyclone Tracy which made landfall in Darwin, Australia in December 1974 with a radius of tropical storm winds of 31 miles (50 kilometers). The relative sizes of these storms, along with the 1995 Labor Day Hurricane which crossed the Florida Keys, the New England Hurricane of 1938, Camille (1969), Gilbert (1988) which made landfall in Mexico, and Andrew (1992) are shown in Figure 3.15. The most intense tropical cyclone on record, as measured by its central pressure, was also Typhoon Tip (870 mb). In the Atlantic Basin, it was Gilbert in 1988 with 888 mb (Lawrence and Gross, 1989; Willoughby, Masters and Landsea 1989). The longest lived tropical cyclone on record was John, which lasted for 31 days (August–September 1994) in the Pacific Ocean. In the Atlantic Ocean Basin, it was was Hurricane Ginger in 1972, which lasted for 30 days.

3.3.3 Tropical cyclones in the Atlantic Ocean Basin

At latitudes around 30°N in the Atlantic Ocean, two large relatively stationary high pressure systems, which extend upward throughout the lower atmosphere (i.e. the troposphere) commonly exist throughout the year. High pressure systems exhibit a clockwise wind flow around their center. These pressure systems are popularly known as the Azores High and the Bermuda High since their centers of highest pressure are often located near those islands. In the Northern Hemisphere, south of these high pressure regions, trade winds blow from east to west. These are called trade winds because they were used by sailing vessels to travel between Europe and the New World prior to the 1800s.

The latitude of the highest pressure of the Azores and Bermuda Highs moves north and south with the seasons, with its northernmost location occurring around August. A lag of about two months with respect to the June solstice (which is when solar radiation is greatest in the Northern Hemisphere) is associated with the time of maximum ocean surface temperatures. This lag occurs because the ocean and the atmosphere take time to warm, analogous to the observation that our warmest temperatures during the day occur after noon (and similarly the warmest temperatures at most locations in the Northern Hemisphere occur in July and August). Thus in the western Atlantic, the ocean is warmest in August and early September.

Tropical storms and hurricanes frequently develop from surface low pressure systems that move westward off the west African coast, called tropical waves (e.g. see Pasch and Avila 1994; Avila and Pasch 1995). The systems move west in the trade wind region. Often such low pressure systems

Midgets

Normal to Large Hurricanes

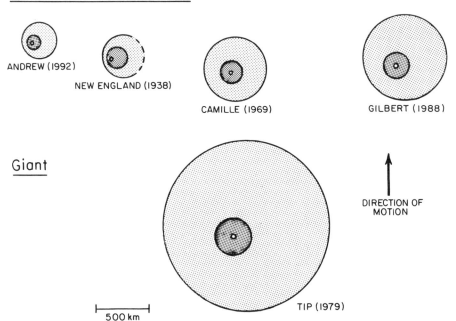

Giant

Figure 3.15   Relative size comparison of the Labor Day Hurricane of 1935, the New England Hurricane of 1938, Hurricane Camille (1969), Typhoon Tracy (1974), Typhoon Tip (1979), Hurricane Gilbert, and Hurricane Andrew (1992). The approximate area with tropical storm shading is shown in light gray; the hurricane wind region is in black. The eye is shown by the white dot (adapted from Chip Guard (1997, personal communication) and McAdie and Rappaport (1991)). The storms are classified in this figure as midgets, normal to large, and giant. The information for the 1938 New England hurricane was provided by E. Boose, 1997 (personal communication) using data from Boose, Foster and Fluet (1994). For the New England Hurricane of 1938, the dashed line on the right side indicates that sustained tropical storm strength winds extended beyond this distance. The information used to define the wind field for Hurricane Andrew in 1992 was obtained from Powell and Houston (1996). Note that in the larger storms the area of hurricane winds is more uncertain, and thus is conservatively underestimated in this figure

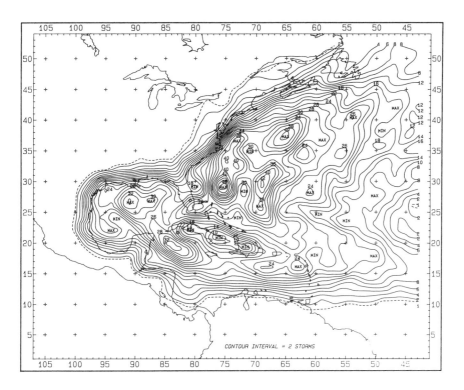

Figure 3.16   This figure shows the number of hurricanes per 100 years passing within about 85 miles (137 km) of any point. Analysis is based on the period of record 1886–1995. Note that in earlier years, storms farther from the coasts were sometimes not detected. Thus the figure is more reliable closer to land. Source: C.J. Neumann, 1997, personal communication. Reproduced by permission of Charles Neumann

recurve northwest and then northeast around the high pressure systems, where they are eventually absorbed into mid-latitude weather systems.

Figure 3.16 shows the probability of a tropical cyclone passing through regions of equal size in the western Atlantic, Gulf of Mexico and Caribbean (Neumann 1996). The highest probabilities are in the lesser Antilles, through the Yucatan straits between the Gulf of Mexico and the Caribbean, and north of the Bahamas east of the southeastern coast of the United States. Since 1961, seven tropical storms or hurricanes in the Atlantic Ocean Basin have been monitored as crossing over into the Pacific Ocean west of Mexico and were tropical storms or hurricanes in that ocean basin. In that time period, one Pacific hurricane crossed over to the Atlantic Ocean basin and became a tropical storm. Tropical weather systems which do not become tropical cyclones in the Atlantic Ocean Basin are the origin of almost all eastern Pacific tropical cyclones (Pasch and Avila 1994; Avila and Pasch 1995).

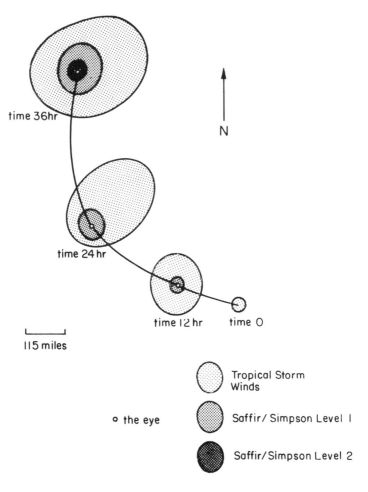

Figure 3.17   Use of the Saffir/Simpson scale to delineate changes in storm intensity over time

Tropical cyclones, of course, are not points or line segments but have areas of influence. A more informative method of presenting movement over time is to display the areas of wind speeds, and the region of different intensities of hurricane force winds, using the Saffir/Simpson scale. Figure 3.17 schematically shows how such a presentation would appear for a storm that evolves from a tropical storm (at time 0) to a hurricane level 2 (at time 36 hours). The National Hurricane Center routinely produces such maps, delineating tropical storms, hurricane, and intense hurricane strengths. Figure 3.18 shows such a map for Hurricane Fran in 1996.

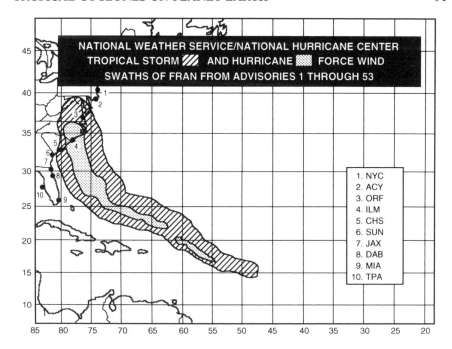

Figure 3.18   National Hurricane Center map for Hurricane Fran in 1996 demarking tropical storm, hurricane, and intense hurricane strength over time. Such maps can be located for all hurricanes at the NC WWW site at http://www.nhc.noaa.gov

The impacts of tropical cyclones are felt to some extent on every continent on the planet. Society is particularly concerned regarding the impacts of intense hurricanes in regions of heavy population density. However, weak hurricanes, tropical storms, and tropical depressions also pose significant threats to society through heavy rain and winds. To respond to these threats, scientists have devoted considerable attention, and policy-makers considerable resources, to understanding and predicting the movement and intensity of tropical cyclones. It is to that subject that we now turn.

CHAPTER 4

Hurricane Forecasts

4.1 TROPICAL CYCLONE MOVEMENT

Tropical cyclones move because the storm is embedded in a larger-scale region of moving air, referred to as the steering current, which tends to move the low-level low pressure center, upper-level high pressure and associated cluster of thunderstorms in the direction of that flow (e.g. see Riehl and Burgner 1950; Riehl and Shafer 1946; Simpson 1946). Tropical cyclones of different intensity are steered by winds at different levels in the troposphere (Figure 4.1).[1] The cyclone itself, of course, is part of the large-scale flow, and its motion is also influenced by its own internal circulation. This sets up a complex process of interaction that is a challenge to predict. Yet accurate prediction of a hurricane's movement is central to short-term decisions to protect life and property.

4.2 EXTERNAL FLOW: THE STEERING CURRENT

If the steering flow were fixed in time, hurricane track forecasting would be comparatively simple. Unfortunately, this is not the case, as the orientation and strength of the steering current changes in response to the position of large-scale pressure features. Contrary to popular conception, however, in the Atlantic most tropical cyclones have fairly regular, well-defined tracks because the location and orientation of the Bermuda and Azores high pressure systems, which determine the track of most Atlantic tropical cyclones, usually change only slowly during the hurricane season. However, the difficulty in predicting a storm track occurs either when the typical climatological steering

---

[1] G.J. Holland of the Bureau of Meteorology in Melbourne, Australia suggests using the winds averaged within a concentric band of 125–250 miles (200–400 km) from the storm center. In another study, tropical cyclones were found to move about 2–4 miles per hour (1–2 meters per second) faster and 10 to 20 degrees to the left of the mean wind flow between about 5000 feet (~1.5 km) and 30 000 feet (~9 km) averaged over an area within a 5 to 7 degree of latitude radius centered on the storm (McElroy 1996).

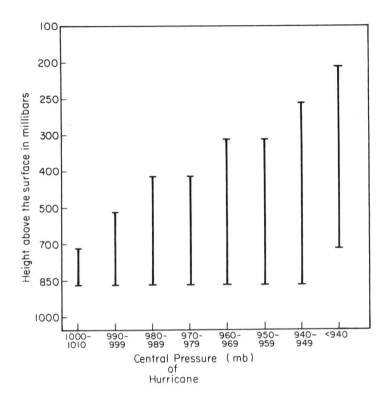

Figure 4.1   The layer of the atmosphere which steers storms of various intensities. Note that weaker storms are steered by a shallow layer of winds lower in the atmosphere and stronger storms are steered by a deeper layer of the atmosphere (adapted from Holland 1993b)

wind flow is replaced by a less common, large-scale flow or, even more importantly, when rapid changes in time occur in the strength and orientation of the steering current.

For example, on 4 September 1965, Hurricane Betsy was moving northwest around the southern rim of the large Bermuda High in the central Atlantic. The track of the storm is shown in Figure 4.2. As the storm was moving northward off the east coast of the United States in a climatologically expected direction and speed, a re-adjustment occurred in the steering current because of a low pressure system associated with a cold front over the central United States. This change resulted in the movement of the Bermuda High towards the west until it was centered north of the storm system. As a result, Hurricane Betsy was blocked from continuing its expected northward movement and became stationary. The Bermuda High center continued to build westward so that, after about a day, the steering currents became northerly and the storm began to move south towards the northern Bahamas. With the re-establishment of the

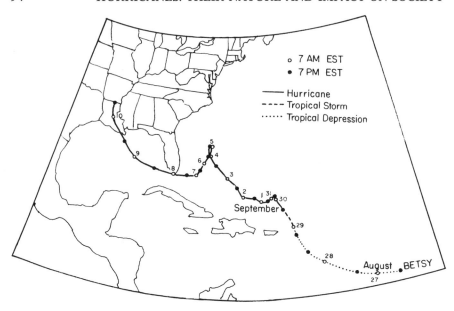

Figure 4.2   Track of Hurricane Betsy from 27 August to 10 September, 1965. Source: DOC (Department of Commerce) 1965

High center to the west, the subsequent track of Betsy traveled around the new position of the Bermuda High, eventually slamming into New Orleans when it finally began once more moving northward around the western flank of the Bermuda High. A major forecast problem associated with this storm was when it would begin its turn towards the west around the southern periphery of the High. An earlier turn would have brought Betsy onshore near Miami, with possible major devastation to that urbanized area. A later turn would have permitted the storm to pass through the Florida Straits. As it happened, the storm crossed over the Florida Keys.

The steering current associated with Hurricane Andrew is shown in Figure 4.3. Displayed in these figures are the wind speeds and directions averaged across the troposphere from a height of about 1 mile (850 mb) to about 7.5 miles (200 mb). The position of Andrew's center at each of the times is superimposed on the figures. The movement of Andrew by this steering current is evident in Figure 4.3 as it traveled westward across South Florida and then northwest into Louisiana.

Figure 4.3   (a–h) Steering current (defined as the average wind speed and direction between 850 millibars and 200 millibars) for Hurricane Andrew at 12-hour intervals starting at 7 am Eastern Standard Time on 22 August 1992. In the figure, the length of the arrows represents the speed of the steering current. A unit arrow of 56 miles per hour (~25 meters per second) is displayed. The position of Andrew's center is shown by the hurricane symbol. (Figure prepared by Joe Eastman, Colorado State University)

(a) 7 AM   AUGUST 22, 1992

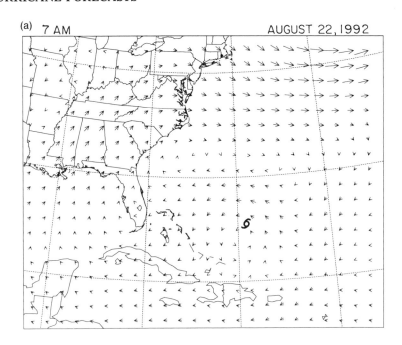

(b) 7 PM   AUGUST 22, 1992

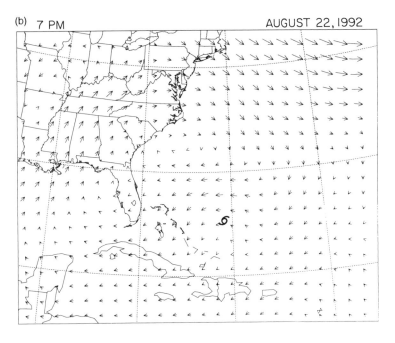

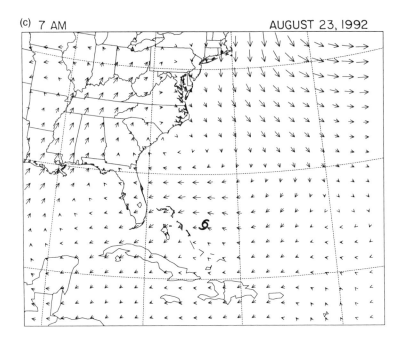

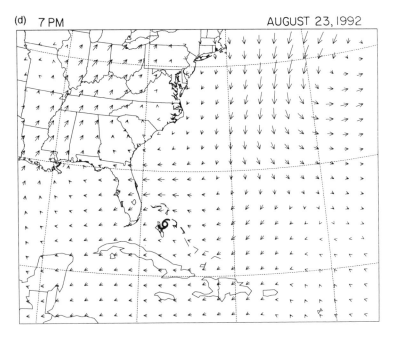

Figure 4.3   (*continued*)

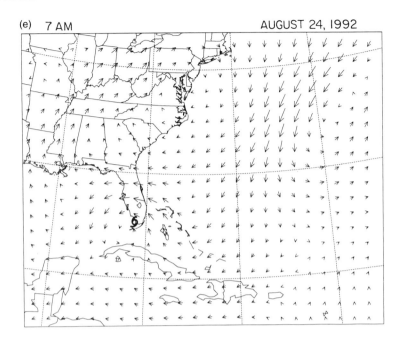

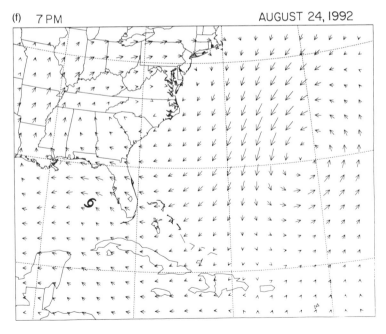

Figure 4.3   (*continued*)

(g) 7 AM   AUGUST 25, 1992

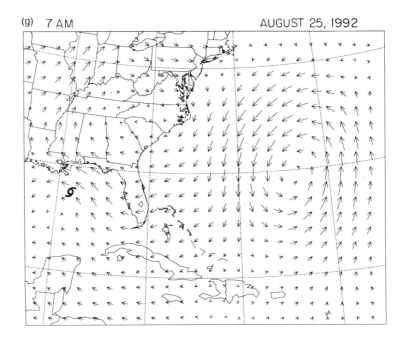

(h) 7 PM   AUGUST 25, 1992

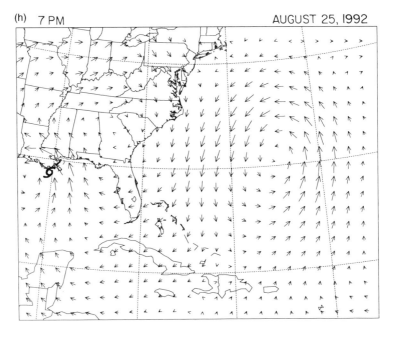

Figure 4.3   (*continued*)

While the news media often attribute hurricanes "with a life of their own", they are, of course, generally well-behaved natural phenomena and, to a large extent, their movement can be explained by the steering currents alone, as shown in the two examples discussed in the last two paragraphs. The difficulty in forecasting their motion occurs when the steering currents are weak and ill-defined and/or when the future state of the steering currents is uncertain. In addition, the need to forecast fairly precise points of landfall to aid emergency planning also contributes to the difficult task of hurricane forecasting.

Tropical cyclones occasionally undergo rapid acceleration in forward motion. This happens when the storm becomes linked to a strong mid-latitude weather system. In addition, tropical cyclones can become absorbed into developing mid-latitude storms thereby infusing added moisture and wind energy from the tropical cyclone and resulting in a more intense mid-latitude storm than otherwise would occur.

An example of a storm that accelerated rapidly out of the tropics was the New England hurricane of 1938. The development of strong, southwesterly winds to the west of this hurricane, associated with a developing mid-latitude storm, resulted in its rapid acceleration to the north at a forward speed of more than 60 mph (27 meters per second). The storm crossed Long Island, New York with little warning, resulting in more than 600 deaths in New England (see Chapter 2).

In 1954, Hurricane Hazel also underwent a similar rapid acceleration to a speed of 60 mph (27 meters per second), as strong south to southwesterly winds developed to the west of the storm. Hazel crossed the North Carolina coastline at 9:25 am on 15 October, and reached Toronto, Canada only 14 hours later where it resulted in 80 deaths (Joe et al. 1995). At that time, it was the most destructive hurricane to reach the North Carolina coast. Every fishing pier was destroyed over a distance of 170 miles (270 km) from Myrtle Beach, South Carolina to Cedar Island, North Carolina. All traces of civilization were practically annihilated at the immediate waterfront between Cape Fear and the South Carolina state line. In 1989, Hurricane Hugo accelerated onto the South Carolina coast at Charleston in association with southeasterly winds caused by a low pressure area in the northeastern Gulf of Mexico, in combination with the Bermuda High to the northeast.

4.3 INTERACTION OF THE STEERING CURRENT AND THE HURRICANE

If the steering current in the immediate vicinity of the storm was constant, its influence on storm motion would be relatively straightforward. Unfortunately, this is generally not the case. The steering current speed and direction are never constant and change both in location and time. If the steering current, for example, becomes stronger towards the right of a hurricane, with respect to its motion, the tendency is for the storm to move towards the right and to slow down. If, however, the steering current becomes weaker towards the right of a moving hurricane, the tendency is for the storm to turn to the left. In addition, a downstream speed-up of the steering current would accelerate a storm, while a downstream deceleration would slow it down (Holland 1983).

The spatial structure of the steering flow directly influences storm motion because the hurricane and steering current are not separate, distinct features but are intertwined with one another. The hurricane is not like a spinning cork flowing down a stream but is more analogous to an eddy within a stream or cream poured into coffee. Just as with a hurricane, an eddy that is rotating counterclockwise tends to move toward a region in which the flow structure enhances the counterclockwise rotation.

The Earth's rotation (as represented by the Coriolis effect; see Figure 3.2) will also influence storm motion. Since the Coriolis effect is greater at higher latitudes, it contributes to the counterclockwise circulation on the west side of a storm (in the northern hemisphere) where northerly winds occur. The net result is a tendency for a poleward and slightly westerly drift with respect to the steering current.

The thunderstorms associated with the hurricane also modify the steering current (Wang and Holland 1996). The air in the upper tropospheric outflow (divergence) descends at some distance from the storm. If this outflow of air were uniform around the storm, there would be no direct effect on the steering current. Observational analysis, however, shows that this outflow is often concentrated in narrow regions in what are called "outflow jets", which can be observed from satellites. If, for instance, this accumulation of air occurred in the front right quadrant with respect to a storm's movement around the southwest side of the Bermuda High, the consequence would be a strengthening of the Bermuda High from what would occur in the absence of the storm (due to the introduction of air to the high pressure system which increases its pressure further). The net result is a movement of the storm to the left of the track that it would have in the absence of this effect. The neglect of this effect in the forecasting of the landfall location of Hurricane Gilbert in 1988 resulted in an erroneous storm track prediction by an independent forecast group of a landfall on the Texas coast. Hurricane Gilbert made final landfall on the northeast coast of Mexico (Sheets 1990). Correspondingly, an outflow jet which results in an accumulation of air in the right rear quadrant of a westward-moving hurricane would tend to accelerate the storm northward in the northern hemisphere.

The importance of outflow jets on storm motion becomes more significant for larger and more intense storms for which the quantity of air in the outflow is greater. It also becomes a more important component in determining the track of a hurricane or tropical storm when the steering currents are weak and ill-defined. Otherwise, when the steering current is strong, outflow jets from

Figure 4.4 Infrared color image of Hurricane Fran taken by a geostationary satellite on 5 September 1996. The image illustrates the cirrus cloud outflow from the upper levels of the storm which appear as wispy white clouds. They are best seen to the north of the storm stretching from West Virginia over New York and east over the Atlantic. Image courtesy of Ray Zehr, NOAA

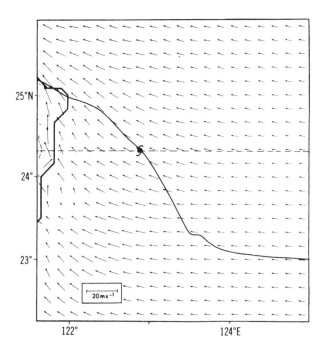

Figure 4.5   Model simulation of the influence of the island of Taiwan on a hurricane track (solid line). (North is to the top of the figure.) Without the presence of the island or the storm, the wind would be easterly at all locations. The island modifies the steering flow such that a southerly steering flow develops just to its east. Source: Bender, Tuleya and Kurihara (1987). Reproduced with permission of the American Meteorological Society

the tropical cyclone are only of secondary importance in terms of determining storm motion.

Figure 4.4 (see color plate) presents a satellite picture to illustrate an outflow jet (Hurricane Fran, 1996). The large avenue of bright clouds stretching to the north and northeast of the storm are cirrus clouds which were transported to that quadrant of the system by the outflow jet. These cirrus clouds are evidence of this air movement to the periphery of the storm through upper-level divergence as thunderstorms reach the upper troposphere. These cirrus clouds will continuously dissipate on their downward edge as the air sinks in this region. The outflow from a tropical cyclone can also counteract the destructive effect on the storm of a vertical shear of the larger-scale wind field, by deflecting those winds around this outflow (Elsberry and Jeffries 1996).

Mountainous islands also influence the movement of a hurricane through the alteration of the steering current. Figure 4.5 illustrates a numerical model simulation of the influence of a large mountainous island on the track of a storm. The winds representing the steering current are easterly in the absence of the island. With the island present to block the flow (even without a hurricane present) the winds would turn southerly to the east of the island. With the hurricane present as shown, the circulation around the storm, interacting with the simulated terrain, resulted in the path plotted in the figure, which is to the right of the steering current in the absence of the storm. In the absence of the island, the hurricane would have moved on a general westward track. In the Atlantic region, terrain effects of this type occur associated with the larger islands of the Caribbean (Cuba, Hispañola). Flatter landscape, such as the Yucatan Peninsula and Florida, and the smaller islands of the Caribbean and Atlantic have much less of an effect on cyclone tracks.

4.4 INTERNAL FLOW

Even when the steering current is relatively uniform and steady, however, hurricane motion is often somewhat irregular. For example, Figure 4.6 illustrates the oscillation of the movement of the eye of Hurricane Dora (1964) as it progressed on a general track westward towards the upper east coast of Florida. This oscillation is primarily due to forces within the hurricane. In Hurricane Anita (1977), the eye and eye wall were observed by aircraft to have an oscillation around the mean track with an amplitude of 3 miles (5 km) and a period of 6 hours (Willoughby 1979). This small-scale irregular behavior of the center of the storm has been attributed to the thunderstorms and strong winds in the eye wall region, which causes the center to deviate short distances to the left or right of its track, similar to the motion of a spinning top. The larger circulation envelope of the hurricane, with its much greater inertia, more closely follows the steering current and tends to force the eye wall back towards the center of the larger circulation.

4.5 TROPICAL CYCLONE TRACK, INTENSITY, AND SEASONAL FORECASTING

Hurricane predictions in the United States are prepared at the National Hurricane Center in Miami, Florida for tropical cyclones in the Gulf of Mexico, the Caribbean and the Atlantic Ocean and the eastern Pacific (Figure 4.7). Other regions of responsibility are shown in Appendix D.

4.5.1 Tropical cyclone track predictions

The National Hurricane Center utilizes a suite of models to forecast tropical cyclone tracks (DeMaria 1995; Aberson and DeMaria 1994). They include one based on climatology and persistence (CLIPER), a statistical model which

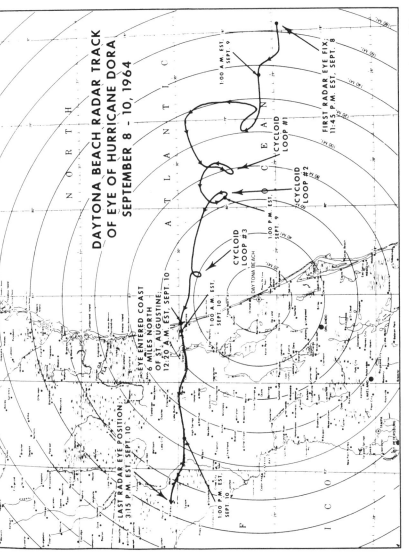

Figure 4.6 Daytona Beach radar track of eye of Hurricane Dora, 8–10 September 1964. The wobbles seen in the Hurricane's motion as it made landfall are not completely understood or predictable. Possible explanations include the influence of land on the storm's circulation, and resultant wobbling similar to a spinning top which is perturbed slightly from its path. Source: DOC (Department of Commerce) 1964

104

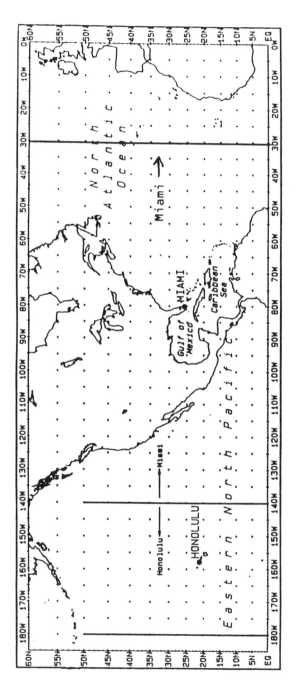

Figure 4.7 Hurricane responsibility area of the National Hurricane Center in Coral Gables, Florida

uses information from the National Weather Service (NWS) global prediction model (NHC 90; McAdie 1991), and several which solve mathematical equations for atmospheric flow including VICBAR (Aberson and DeMaria 1994), BAM (Marks 1992), the Geophysical Fluid Dynamics Lab (GFDL) model (Bender et al. 1993), and the National Meteorological Center (NMC) Aviation Medium Range Forecast (MRF) global forecast models (Lord 1993). During the 1992 and 1993 seasons, the GFDL model track forecasts were superior to the other track models (Aberson and DeMaria 1994); the improved forecasts of hurricane track using this model are also summarized by Sawyer (1995).

Track forecasts are very sensitive to how the actual hurricane is initially defined in the model (Leslie and Holland 1995). For instance, track predictions depend on the choice of hurricane radius and strength, and its initial location. Improved measurements of temperature, wind, and humidity in and around tropical cyclones have contributed to improved operational track forecasts (Burpee et al. 1996). An example of a 72-hour track forecast for Hurricane Hugo for different models is given in Figure 4.8. This figure illustrates that despite significant progress in hurricane forecasting, exact track prediction remains fraught with difficulties. Hugo actually made landfall near Charleston, South Carolina. Hurricane track models are also summarized in Puri and Holland (1993).*

Figure 4.9 illustrates the trend and accuracy of 24-hour, 48-hour, and 72-hour forecasts of storm position between 1970 and 1992. As of 1997, average forecast errors are on the order of 115 miles (185 km) for 24-hour forecasts, 230 miles (368 km) for 48-hour forecasts, and 345 miles (552 km) for 72-hour forecasts (C. Landsea, 1997, personal communication). Note that an improvement of only about 20 miles (42 km) has been achieved in 24-hour position forecasts over 23 years, despite the great advances both in monitoring these storms (e.g. radar, satellite, reconnaissance aircraft) and in computer power to process and analyze the data.

---

* Four criteria have been proposed for accurate track forecasts using models such as VICBAR, BAM, the GFDL model, and the Aviation global model (Elsberry 1995). These are: adequate initial specifications of the environmental wind field, the symmetric and asymmetric cyclone vortex structure, and the adequacy of the prediction models to forecast the time evolution of the vertical and horizontal wind field. An accurate representation of the diabatic heating of the atmosphere by the hurricane, and the prediction of winds and temperature in the upper troposphere are also essential to accurately characterize hurricane–environmental interactions (Wu and Kurihara 1996). When the large-scale weather pattern is changing with time, these requirements are difficult to achieve with sufficient accuracy.

The MRF model is used for general weather forecasting in addition to its application for tropical cyclone track prediction. VICBAR, BAM, NHC90 and the GFDL model use forecast fields from the MRF model for input. One version of NHC90 (referred to as UK90) uses output from the United Kingdom Meteorological Office global forecast model.

The GFDL model includes the most physical realism in its formulation, including moving nested grids which translate with the cyclone and a sophisticated vortex initialization scheme with the finest horizontal grid interval of 20 km (DeMaria 1995; Bender et al. 1993).

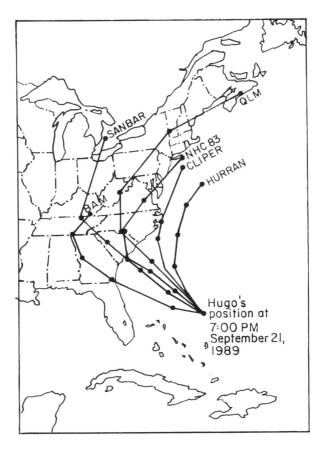

Figure 4.8   National Hurricane Center 72-hour forecast tracks for Hurricane Hugo 1989, starting from its position at 0 GMT on 21 September 1989. The forecast 12-hour positions for each of six forecast techniques are shown as black dots

Since 1983, probabilities of a tropical cyclone passing within 75 miles (121 km) of specific geographic locations have also been publicly distributed. An example of the format used in these probability forecasts is shown in Figure 4.10, in this case for Hurricane Erin in 1995.

4.5.2 Tropical cyclone intensity change predictions

Forecasting of changes in tropical cyclone intensity is a much more difficult task than forecasting tropical cyclone tracks. Several methods are used to predict changes in intensity. The simple climatology and persistence intensity technique (SHIFOR) and the statistical hurricane intensity prediction scheme

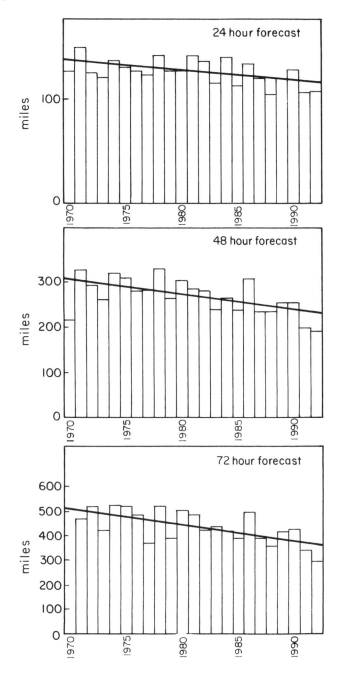

Figure 4.9 Annual average NHC official forecast errors for 24 hours, 48 hours, and 72 hours (1970–1992). The diagonal line shows the trend. Source: McAdie and Lawrence (1993)

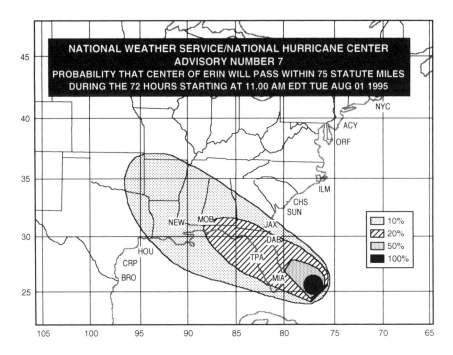

Figure 4.10   Forecast probability that the center of Erin will pass within 75 miles during
the 72 hours starting at 11 am Eastern Daylight Time on 1 August 1995 (redrawn from
the National Hurricane Center WWW page at http://www.nhc.noaa.gov)

(SHIPS) are used for tropical cyclone intensity forecasts. SHIFOR, analogous
to the CLIPER track model, uses only climatology, persistence, and current
storm characteristics to calculate statistically the most expected intensity
change. SHIPS, in contrast, utilizes selected current meteorological and ocean
data, including the difference between the current tropical cyclone intensity
and its potential maximum based on sea surface temperature (DeMaria and
Kaplan 1994). SHIPS has average intensity errors that are 10–15% smaller
than those for SHIFOR. The GFDL model, briefly described in Section 4.3.1,
also predicts intensity change. However, its horizontal resolution is insufficient
to resolve the eye wall region, which is critically involved with the intensification process. Preliminary research modeling with finer spatial resolution
for Hurricane Andrew suggests that the more detailed representation of the
thunderstorms in the eye wall region and in the inflow to the hurricane can
lead to improved intensity change forecasts (Eastman 1995, Lin et al. 1997).
Operational techniques have so far shown little skill in intensity change
prediction; thus forecasters primarily rely on empirical techniques.

Satellite imagery is often used to estimate the intensity of clusters of oceanic
tropical thunderstorms, including tropical cyclones and hurricanes. This

approach is particularly useful when reconnaissance aircraft are unavailable to monitor the strength of such a thunderstorm cluster. The use of satellite images for this purpose is based on a pattern recognition decision tree (Dvorak 1975; 1984). The difference in the temperature of the eye of the hurricane, and the cloud top temperatures of the surrounding eye wall, based on infrared satellite images, is one example of the use of this technique.

When tropical systems move close enough to land, aircraft reconnaissance is also used to monitor their intensities (Neal Dorst in Landsea 1997). These flights are conducted by the US Air Force Reserve 53rd Weather Reconnaissance Squadron and the National Oceanographic and Atmospheric Administration (NOAA) Aircraft Operations Center. The 10 WC-130 Air Force planes of the squadron are based at Keesler Air Force Base in Mississippi, but, as needed, are positioned elsewhere, including islands in the the eastern Caribbean Sea. Measurements include wind, temperature and humidity at flight level, as well as data collected by dropping instruments along the flight path. The three NOAA aircraft (two P-3 Orions and a Gulfstream IV), based at MacDill Air Force Base in Florida have more sophisticated instrumentation, including on-board weather radar. These three aircraft are generally used only for hurricanes that are threatening landfall, or otherwise have specific scientific interest.

4.5.3 Seasonal predictions of tropical cyclone activity

Researchers also prepare forecasts for entire seasons. Professor William Gray of Colorado State University leads a team that forecasts tropical cyclone activity for upcoming seasons in the Atlantic, Gulf of Mexico and Caribbean (Gray et al. 1995). Table 4.1 summarizes Gray's forecasts from 1984 to mid-1997.

Their forecasts, also made in early August for the remainder of the tropical cyclone season, are based on a number of factors including (Gray 1995; Landsea et al. 1994):

1. The winds at a height of about 15 miles (about 24 kilometers) and about 13 miles (about 21 kilometers). There is increased hurricane activity when the winds are more westerly than average and where there are smaller differences in wind between the two levels. These winds fluctuate between east and west in a cycle that is slightly longer than two years and is called the "stratospheric quasibiennial oscillation". Thus this factor alone would tend to make seasons vary between active and quiet from one year to the next (Shapiro 1989).
2. The state of the El Niño–Southern Oscillation (ENSO) cycle. A warm event in the equatorial East Pacific is associated with reduced hurricane activity, while a cold event is associated with enhanced activity (on ENSO see Glantz 1996).

Table 4.1 Seasonal forecasts of Atlantic hurricane activity produced by Professor William Gray of Colorado State University, and actual hurricane incidence. Forecasts available at http://tropical.atmos.colostate.edu/forecasts/.

Forecast of	1950–1993 Mean	1984	1985	1986	1987	1988	1989	1990	1991	1992	1993	1994	1995	1996	1997
A. November (9 to 14 month)															
No. of hurricanes	5.7									**4** *4*	**6** *4*	**6** *3*	**8** *11*	**5** *9*	**6** –
No. of named storms	9.3									**8** *6*	**11** *8*	**10** *7*	**12** *19*	**8** *13*	**10** –
No. of hurricane days	23									**15** *16*	**25** *10*	**25** *7*	**35** *62*	**20** *45*	**24** –
No. of named storm days	46									**35** *38*	**55** *30*	**60** *28*	**65** *121*	**40** *78*	**50** –
Hurricane destruction potential	68									**35** *51*	**75** *23*	**85** *15*	**100** *173*	**50** *135*	**60** –
No. of Category 3-4-5 hurricanes	2.2									**1** *1*	**3** *1*	**2** *0*	**3** *5*	**2** *6*	**2** –
Category 3-4-4 hurricane days	4.5									**2** *3.25*	**7** *0.75*	**7** *0.0*	**8** *11.5*	**5** *13*	**4** –
B. Late May/early June (0 to 6 month)															
No. of hurricanes	5.7	**7** *5*	**8** *7*	**4** *4*	**5** *3*	**7** *5*	**4** *7*	**7** *8*	**4** *4*	**4** *4*	**7** *4*	**5** *3*	**8** *11*	**6** *9*	**7** –
No. of named storms	9.3	**10** *12*	**11** *11*	**8** *6*	**8** *7*	**11** *12*	**7** *11*	**11** *14*	**8** *8*	**8** *6*	**11** *8*	**9** *7*	**12** *19*	**10** *13*	**11** –
No. of hurricane days	23	**30** *18*	**35** *21*	**15** *10*	**20** *5*	**30** *24*	**15** *32*	**30** *27*	**15** *8*	**15** *16*	**25** *10*	**15** *7*	**35** *62*	**20** *45*	**25** –
No. of named storm days	46	**45** *51*	**55** *51*	**35** *23*	**40** *37*	**50** *47*	**30** *66*	**55** *68*	**35** *22*	**35** *38*	**55** *30*	**40** *28*	**65** *121*	**45** *78*	**55** –
Hurricane destruction potential	68					**75** *81*	**40** *108*	**90** *57*	**40** *23*	**35** *51*	**65** *23*	**35** *15*	**110** *173*	**60** *135*	**75** –
No. of Category 3-4-5 hurricanes	2.2							**3** *1*	**1** *2*	**1** *1*	**2** *1*	**1** *0*	**3** *5*	**2** *6*	**3** –
Category 3-4-5 hurricane days	4.5								**2** *1.25*	**2** *3.25*	**7** *0.75*	**7** *0.0*	**6** *11.5*	**5** *13*	**5** –
C. Late July/early August (0 to 4 month)															
No. of hurricanes	5.7		**7** *7*	**4** *4*	**4** *3*	**7** *5*	**4** *7*	**6** *8*	**3** *4*	**4** *4*	**6** *4*	**4** *3*	**9** *11*	**7** *9*	**7** –
No. of named storms	9.3		**10** *11*	**7** *6*	**7** *7*	**11** *12*	**9** *11*	**8** *14*	**7** *8*	**8** *6*	**10** *8*	**7** *7*	**16** *19*	**11** *13*	**9** –
No. of hurricane days	23		**30** *21*	**10** *10*	**15** *5*	**30** *24*	**15** *32*	**25** *27*	**10** *8*	**15** *16*	**25** *10*	**12** *7*	**30** *62*	**25** *45*	**45** –
No. of named storm days	46		**50** *51*	**25** *23*	**35** *37*	**50** *47*	**35** *66*	**50** *68*	**30** *22*	**35** *38*	**50** *30*	**30** *28*	**65** *121*	**50** *78*	**78** –
Hurricane destruction potential	68					**75** *81*	**40** *108*	**75** *57*	**30** *23*	**35** *51*	**50** *23*	**30** *15*	**90** *173*	**70** *135*	**70** –
No. of Category 3-4-5 hurricanes	2.2							**2** *1*	**1** *2*	**1** *1*	**2** *1*	**1** *0*	**3** *5*	**3** *6*	**3** –
Category 3-4-5 hurricane days	4.5							**5** *1.0*	**1** *1.25*	**2** *3.25*	**7** *0.75*	**7** *0.0*	**6** *11.5*	**5** *13*	**5** –

Blank areas in the table indicate that these forecasts were not made (Gray 1994; 1995). These forecasts are available at http://tropical.atmos.colostate.edu/forecasts/index.html.

Forecasts are given in bold type, actual values in italic.

Named storm: A tropical storm or hurricane.

Hurricane day: Four 6-hour periods during which a tropical cyclone is estimated to have hurricane-strength winds.

Named storm day: Four 6-hour periods during which a tropical cyclone is observed or estimated to have tropical storm or hurricane-strength winds.

Hurricane destruction potential: A measure of a hurricane's potential for wind and storm surge destruction defined as the sum of the square of a hurricane's maximum wind speed (in 10^4 knots^2) for each 6-hour period of its existence.

Category 3-4-5 hurricane day: Four 6-hour periods during which a tropical cyclone has intensity of Saffir/Simpson Category 3 or higher.

3. Prior rainfall in the western Sahel region of Africa. Wetter conditions are related to increased hurricane activity.
4. Spatial differences of surface pressure and surface temperature over western Africa during the previous February through May. Large differences correspond with more hurricane activity.
5. Average sea level pressure in the Caribbean in the previous spring and early summer. If the pressures are low, hurricane activity is generally enhanced.
6. Winds at a height of about 7.5 miles (about 12 km) for five low latitude sites in the Caribbean. Hurricane activity is increased if the easterly winds are stronger than average.
7. Warmer sea surface temperatures in the North Atlantic help create more Atlantic hurricanes. This factor has been used in the seasonal hurricane forecasts since August 1996 (C. Landsea, 1997, personal communication).

The seasonal forecast is for named tropical cyclones, hurricanes, intense hurricanes (Category 3 or stronger storms); the number of days with tropical cyclones, hurricanes, and intense hurricanes; hurricane destruction potential; and net tropical cyclone activity. Over a 45-year period (1950–1994), there were annual averages of 2.1 intense hurricanes and 4.5 intense hurricane days. During this period the maximum number of intense hurricanes was seven in 1950, while the largest number of intense hurricane days was 21 in 1961.

The seasonal forecasts indicate that above-average hurricane seasons occur when the Sahel region is wetter. In this situation, areas of thunderstorms that propagate across this area of Africa (and are a major source for tropical cyclones in the Atlantic, Gulf of Mexico, and Caribbean; Pasch and Avila 1994) are in a more favorable environment since transpiring vegetation, which has grown in response to the rains earlier in the year, provides a water vapor source to the clouds. Thus these systems are more likely to persist when they move westward off the African west coast. The rainfall in west Africa also directly relates to the vertical wind shear in the Atlantic Ocean tropical cyclone development region, with wetter years having lower shear (Goldenberg and Shapiro 1996).

Several of the other predictors relate to a reduction in the vertical shear of the wind as being associated with enhanced hurricane activity. As discussed in Chapter 3, weak shear is a favorable condition for tropical cyclone development. Lower surface pressures in the Caribbean, also an environmental condition that favors development, is associated with convergence of water vapor and heat into propagating thunderstorm cloud clusters, as well as tropical cyclones after they develop.

4.5.4 Attempts at tropical cyclone modification

There have been attempts by scientists to modify the intensity of tropical cyclones. The main hypothesis is that by seeding cumulus clouds with silver

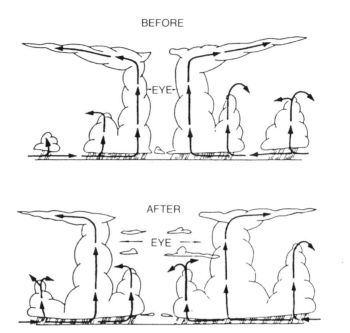

Figure 4.11   Hypothesized vertical cross-sections through a hurricane eye wall and rain bands before and after seeding. Dynamic growth of seeded clouds in the inner rain bands provides new conduits for conducting mass to the outflow layer and causes decay of the old eye wall. Source: Simpson et al. (1978)

iodide outside the eye wall, clouds with liquid water colder than 0°C (referred to as "supercooled water") could be converted to ice crystals. This change of phase of water would release the heat of fusion, thereby enhancing the growth of the cumulus clouds and establishing the eye wall at a greater radius from the center of the storm circulation. Without the silver iodide, it is hypothesized that the cloud droplets would remain liquid. Just as ice skaters slow their rotation when their arms are spread out, the hypothesis is that a larger radius of the eye wall will cause a reduction in wind strength. The hypothesis is sketched schematically in Figure 4.11.

This program of tropical storm modification was called Project Stormfury. Robert and Joanne Simpson were the original source of the Stormfury hypothesis in 1960. Their insight was inspired by an observation of Hurricane Donna (1960) by Professor Herbert Riehl, who noted that nearly all of the outflow cloudiness stemmed from an aggregation of thunderstorms in the front right quadrant of the eye wall. Hurricanes Esther (1961), Beulah (1963), and Debbie (1969) were seeded as part of this project, although only the Debbie experiments closely followed the most recent Stormfury hypotheses.

In an earlier experiment, a hurricane was seeded on 13 October 1947 off the southeast United States coast. Because it subsequently moved westward into the Georgia coast, questions were raised among critics of weather modification as to whether the seeding caused the abrupt change in storm track. More recent analysis strongly suggests that the alteration in direction would have occurred in any case. Hurricane Ginger was also seeded in 1971. However, it was an anomalous storm with an eye wall usually below 20 000 feet (6100 meters) and no significant quantities of supercooled water were found.

Hurricane modification ended because of equivocal research findings including evidence that hurricanes may only infrequently be amenable to modification because of their very large variability, questions of national and international legal issues, and a general loss of support for weather modification efforts (Cotton and Pielke 1995; Pielke and Glantz 1995). Currently, further seeding experiments are not being performed. Project Stormfury was terminated in 1983 (Willoughby et al. 1985). A policy consequence of the inability to control hurricanes is that impacts must be mitigated by reducing societal exposure and *not* by modifying event incidence.

4.5.5 Value to society of forecasts

The production of a hurricane forecast is only the first step in the process of its effective use by decision-makers. A hurricane forecast must also be communicated through a process of – hear, understand, believe, personalize, respond, and confirm (as described by Sorensen and Mileti 1988). A decision by an individual to act in response to the information about an approaching storm, or any other extreme event for that matter, is conditioned by a wide range of factors such as perceived risk, education, time to impact, and many more (Sorensen 1993). Social scientists have a well developed understanding of the process of natural hazards warnings and human response (see, e.g., Drabek 1986). The use of a forecast must be understood in the broad context of a process from production through communication through response.

The US National Weather Service (NWS) produces and issues a number of products related to hurricanes: hurricane watches and warnings, flash flood watches, flash flood warnings, flash flood statements, tornado watches and warnings, severe thunderstorm watches and warnings, severe weather statements, and special weather statements (NOAA 1994). Once a forecast or warning is produced it is typically distributed to a range of Federal, state, and local agencies, including FEMA, the Corps of Engineers, and state and local emergency management agencies. Similarly, the media – including print, television, and radio – also receive NHC weather products and are often a critical link in the communication process. Recently, the World Wide Web has provided a wide range of information resources in both text and graphic formats (http://www.nws.noaa.gov/).

A forecast is only useful if it can be incorporated into the decision process of a particular user. For this to happen, "it is vital that [official] personnel share knowledge of dissemination systems, procedures, capabilities, and response requirements with the media; Federal, state, and local agencies; and the general public involved in the total warning process" (NOAA 1994). Similarly, officials must also be aware of the information needs of and decision-making constraints on users of their information. Effective use of a forecast is *not* achieved simply by sending information from one party to another; it is the result of two-way interaction between the sender and receiver.

A focus on the process of decision-making is central to realistic determination of the opportunities for and limitations on use of forecasts to reduce societal vulnerabilities to hurricanes (Pielke 1994). To focus on process is to focus on the formulation, promulgation, and execution of particular decisions (Lasswell 1971). Often, both scientists and policy-makers alike behave as if the development of scientific information (such as a forecast) is sufficient to lead to better decisions. From a decision-maker's perspective, a call for better information from scientists can forestall the need to make difficult decisions while placing the burden of problem-solving upon the scientists (Clark and Majone 1985). From a scientist's perspective, a focus on information allows for relative autonomy from the "politics" of decision-making and a justification for continued funding. However, it is often the case that scientific information is misused or not used at all because of rigidities in and practicalities of decision-making processes (e.g. Feldman and March 1981).

Specific decisions are made in the context of a set of alternative courses of action. For example, in order to better prepare for the hurricane threat, citizens of New Orleans might desire a range of alternative responses to the following questions: How strictly shall we enforce our buildings codes? At what point in time before an approaching hurricane shall evacuation become mandatory? How much, if any, increase in insurance should citizens be required to carry in the face of a long-term forecast of increased hurricane incidence? Alternative actions in response to each question are embedded in a broader context of values, feasibility, and efficacy. For instance, building code enforcement cannot be separated from issues such as the costs (to the resident) of building a reinforced structure and the tax revenues necessary to hire a sufficient number of building inspectors. In many respects, decisions to which forecasts may be relevant are decisions about how a community wishes to move into the future.

Policy-makers require some sense of the usefulness of short- and long-term forecasts with respect to the hurricane problem in order to determine the amount of resources to be placed into their development versus alternative responses. However, demonstration of use or value of hurricane forecasts is a challenging analytical task. Glantz (1986) notes that "one could effectively argue that the value of climate-related forecasts will in most instances be at least as much a function of the political, economic, and social settings in which they are issued than of the soundness of information in the forecast itself". Put another way, the solution to the hurricane problem is potentially very different in Dade County, Florida, from that in Worcester County, Maryland, and both of those may be significantly different from the solution in Nueces County, Texas as a result of economic, political, and civic differences between the various communities. Further, what works in the mainland United States may not work as effectively on islands or in other countries.

Apart from demonstrating the value of improved forecasts, accurate assessment of societal vulnerabilities to hurricanes is a very challenging task. An example of the difficulties in defining the extent and magnitude of the hurricane threat is provided by the response of the insurance industry to Hurricane Andrew. The 1992 event served as a "wake-up call" to the insurance industry. Prior to Andrew the insurance industry largely ignored hurricane climatology and instead kept records of hurricane-related deaths and economic damage, according to Russell Mulder, director of risk engineering at the Zurich-American Insurance group (Wamsted 1993). The insurance industry's records were accurate measures of their losses, but not of hurricanes: they neglected storms that did not make landfall and underestimated the potential impact of storms that made landfall in relatively unpopulated areas. Since Hurricane Andrew, the insurance industry has paid closer attention to the hurricane threat (e.g. Banham 1993; Noonan 1993; Wilson 1994). One would expect the insurance industry to be among the most sensitive to societal vulnerability to hurricanes; however, Hurricane Andrew demonstrated that even when concern exists, accurate definition of the hurricane problem is difficult.

One expert in the value of forecasts states that "forecasts possess no intrinsic value. They acquire value through their ability to influence the decisions made by the users of the forecasts" (Murphy 1993). Yet, because numerous factors contribute to any particular decision "assessing the economic value of forecasts is not a straightforward task" (Murphy 1994). That is, a forecast is, at best, only one of a multitude of factors which influence a particular (potential) user. It is often difficult to identify the signal of the forecast in the noise of the decision-making process. Factors external to the forecast may hinder its use.

Two complementary approaches to assessment of forecast value can be summarized as use-in-theory and use-in-practice Murphy (1994) calls these *prescriptive* and *descriptive* assessments of forecast value, while Glantz (1977) uses the terminology of "what ought to be" and "what is". Use-in-theory refers to efforts to estimate the "value of forecasts under the assumption that the decision maker follows an optimal strategy" (Stewart 1997). Generally, economists, statisticians, and decision theorists share expertise in assessment of use-in-theory (e.g. Winkler and Murphy 1985). Use-in-practice refers to efforts, including case studies, to understand how decisions are actually made in the real world and the value of forecast information therein (e.g. McNew et al. 1991). Political scientists, sociologists, and psychologists are examples of those with expertise in assessment of use-in-practice.

It is likely that, as forecasts of hurricane incidence demonstrate increased skill, the value of such forecasts will not be self-evident to most users. Hence, it may be worthwhile for producers of both short- and long-term forecasts to conduct an ongoing parallel research effort targeted at actual and potential users. Such a parallel program could focus on assessments of use-in-theory and use-in-practice in order to identify opportunities for and constraints on improved and proper use of hurricane forecasts. Counties, states, and SLOSH basins would be appropriate levels of analysis for an assessment. Particular decisions could be identified from a decision process map, such as that created by Lee County Florida and reproduced as Appendix E. Such assessments may find that in some cases a particular decision process may constrain effective use of a forecast. Other assessments may find clear opportunities to leverage forecast information for reduced vulnerability. If public funding dedicated to the development of improved forecasts of hurricane activity are justified in terms of their value added to social processes, then the sustainability of support for such research may depend in large part upon demonstration of actual use or value.

Where society ought to spend its limited resources to best address the hurricane problem is not clear. The lessons of the weather modification experience provide a warning to the scientific community. Modification of hurricane incidence will remain impractical for the foreseeable future, in spite of the mid-century optimism following an intensive series of efforts to "tame" hurricanes in the 1950s and 1960s (Gentry 1974). Experience with hurricane modification does provide one very important lesson: Care must be taken not to "over-promise" expected benefits deriving from research (e.g. Tennekes 1990; Namias 1980).

Consider the following statement made in the late 1940s in a talk given by Nobel Laureate Irving Langmuir at the dawn of optimism about hurricane modification: "The stakes are large and with increased knowledge, *I think that we should be able to abolish the evil effects of these hurricanes*" (quoted in Byers 1974, emphasis added). On one level such claims reflect the eternal optimism of science and technology. But at another level, such claims are publicly irresponsible and potentially damaging to the institution of science (Changnon 1975). One can easily imagine a policy maker, excited by the possibilities of Langmuir's claim, making an argument that "preparedness plans for hurricanes would no longer be necessary because weather modification scientists had discovered a magic bullet". Of course, taking the thought a step further, had a hurricane then hit a poorly prepared community, it is reasonable to expect that blame would have been laid at the feet of the scientist, and not the policy maker. In the context of forecasts of hurricane activity, credibility with the public will be difficult to gain, and easy to lose (Slovic 1993).

Weather modification is perhaps an extreme example of the risks involved with overselling science. However, in an era when science is increasingly called upon to contribute to the resolution of many difficult societal problems, demonstration of benefits may become central to sustained federal support of research to develop improved forecasting capabilities.

CHAPTER 5

Hurricane Impacts

When a hurricane forms, it poses a significant danger to society. The importance and danger of tropical cyclones differ between land and water. Over the oceans, the human activities and assets at risk are primarily oil rigs, shipping, and air traffic. On land, particularly along the coast, cities, towns, and industrial activities become threatened. Hurricanes also have ecological impacts (e.g. Lodge 1994). This chapter overviews the physical impacts of a hurricane over the ocean, at landfall, and inland. It also discusses societal impacts and how they are measured and understood.

5.1 OCEAN IMPACTS

Winds of hurricane speed over the ocean are characterized by blowing spray over a white, foaming sea state of large waves. Monstrous waves can develop in this environment. For example, in 1995, the cruise ship Queen Elizabeth II was rocked by a 70 foot (21 meter) wave caused by distant Hurricane Luis. The sea near a hurricane is chaotic, and an extreme hazard to shipping can occur in response to wave motion moving in many directions. The heights of the maximum ocean waves that were associated with Hurricane Kate in the eastern Gulf of Mexico in November 1985 are illustrated in Figure 5.1, along with other observed meteorological characteristics. These measurements were made from an unmanned ocean buoy. Wave height was determined by monitoring the movement of the buoy by the waves. In this storm, wave heights of greater than 38 feet (12 meters) were observed preceding the passage of the eye on 20 November.

For comparison, strong winds, of course, also occur in winter storms over the open ocean. The risk to shipping and other activities from wave action, however, is generally less serious in such storms for two reasons. First, the wind blows primarily in one direction in a given sector of a winter storm. Hence the waves move in concert with the wind. A ship can thus orient itself to minimize the effect of the waves. In a hurricane, winds change direction rapidly around the eye. The result is a chaotic sea with swells and waves propagating in a myriad of directions. A ship cannot simply steer into the running sea in order to

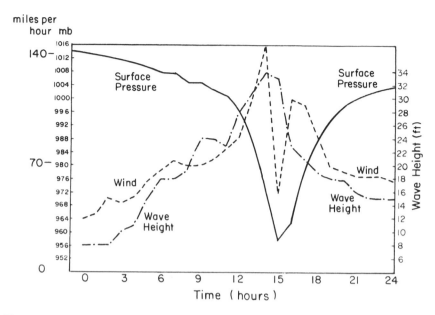

Figure 5.1   Observations of wind speed, wave height, and surface pressure of the passage of Hurricane Kate over 24 hours beginning on 20 November 1985, as monitored by a floating oceanic buoy at 26°N and 86°W. Original figure provided by R.H. Johnson

reduce its risk since there is no one direction from which the waves come. Large waves also superimpose on top of each other, producing enormous swells.

A second important difference between strong winter storms and hurricanes is that the winter storms seldom have winds which exceed the minimal hurricane strength. In contrast, sustained wind speeds up to 200 mph (90 meters per second) are occasionally observed in hurricanes. In 1944, in the western North Pacific during World War II, when an American naval convoy was struck without warning by a Pacific typhoon, three destroyers capsized and sank, nine other ships sustained serious damage, and 19 other vessels received lesser damage (Kotsch 1977).

5.2 LAND IMPACTS AT THE COAST AND A SHORT DISTANCE INLAND

At the coast, the major impacts of either a landfalling hurricane or one paralleling the coast are:

- Storm surge
- Winds

- Rainfall
- Tornadoes.

Of these weather features, the storm surge has accounted for over 90% of the deaths in a hurricane. In recent years, and particularly in the aftermath of Hurricane Andrew, more attention has been paid to the effects of hurricane winds.

5.2.1 Storm surge

"Storm surge" refers to a rapid rise of sea level that occurs as a storm approaches a coastline. This is in addition to changes in variations in sea level due to tides. Thus, a storm surge causes greatest inundation at high tide. A very strong hurricane may produce a storm surge of 20 feet (6 m) of which about 3 feet (1 m) is due to the lower atmospheric pressure at the center of a hurricane. The remaining storm surge is due to the following two factors: (i) the piling up of water at the coast, generated by the strong onshore winds; and (ii) a decreased ocean depth near the coast, which steepens the surge. A common misconception is that the lower pressure at the center of a storm is the primary cause of the storm surge.

At landfall, storm surge is highest in the front right quadrant (with respect to the storm's motion) of a Northern Hemisphere tropical cyclone, where the onshore winds are the strongest. It is also large where ocean bottom bathymetry focuses the wave energy (e.g. as in a narrowing embayment). Peak storm surge from a landfalling cyclone increases with greater wind speeds and the areal extent of the storm's maximum winds, out to about 30 miles (48 kilometers).

Storm surge also occurs when a storm parallels the coast without making landfall (Xie and Pietrafesa 1995). The storm surge will precede the passage of the storm's center when winds blow onshore preceding passage of the eye. Similarly, the surge will lag the storm's center when the hurricane is moving such that onshore winds follow passage of the eye. Offshore winds which are associated with a storm can produce a negative surge as the sea level is lowered by the strong winds blowing out from the coast.

Storm surge is generally estimated to diminish in depth by 1–2 feet (0.3–0.6 meters) for every mile (1.6 kilometers) that it moves inland. Even if the inland elevation were only 4–6 feet (1.2–1.8 meters) above mean sea level, a storm surge of 20 feet (6 meters) might typically reach no more than 7–10 miles (11–16 kilometers) inland. Thus, the most destructive effect of the storm surge hazard is on beaches and offshore islands. Stone et al. (1996) provide a recent evaluation of extensive beach profile changes near Pensacola, Florida as a result of the storm surge associated with the landfall of Hurricane Opal in 1995.

5.2.2 Storm surge analysis

A storm surge can be deadly. In 1900, up to 12 000 deaths occurred in Galveston, Texas, primarily as a result of the storm surge that was associated with a Gulf of Mexico hurricane (Chapter 1). In 1957, a storm surge, which was associated with Hurricane Audrey and which was over 12 feet (3.5 meters) and extended as far inland as 25 miles (40 kilometers) in this particularly low-lying region, was the major cause of the death of 390 people in Louisiana. In September 1928, the waters of Lake Okeechobee, driven by hurricane winds, overflowed the banks of the lake and were the main cause of more than 1800 deaths (Chapter 1).

Areas to be evacuated due to storm surge in the case of hurricane landfall are determined through a model developed by the National Weather Service (NWS) called SLOSH (Sea, Lake, and Overland Surges from Hurricanes; Jarvinen and Lawrence 1985). The SLOSH model is used to define flood-prone areas in 31 "SLOSH basins" along the US Gulf of Mexico and Atlantic coasts (Figure 5.2). Figure 5.3a shows Charleston, South Carolina, as a close-up example of a SLOSH basin. Determination of storm surge vulnerabilities is the result of an interagency and intergovernmental process funded by NOAA, FEMA, Army Corps of Engineers, and various state and local governments (BTFFDR 1995). From development through application the SLOSH process takes about two years. Because coastlines are constantly changing due to human and natural forces, the SLOSH process is never completed.

A SLOSH model is developed through a number of steps for each basin. First, the Techniques Development Laboratory (TDL) of the US National Weather Service (NWS) obtains water depth and land height data for a particular basin to input to the model. Second, the model is tested against historical analogues in the basin, with model runs compared against empirical data from past storms. Third, TDL, the National Hurricane Center (NHC), and local officials refine the water depth and land height data through reconnaissance of the basin. They check for barriers such as highways in order to improve the accuracy of the input data. After final testing for accuracy, each SLOSH model is declared "operational" and is turned over to the NHC for applications.

Evacuation strategies based on the SLOSH model output are generally developed in three phases. First, the NHC simulates between 250 and 500 hypothetical storms in each basin to develop a composite map of potential storm surge inundation for various families of storms (based on strength, movement, and point of landfall). A composite map (called a Maximum Envelope Of Water, or MEOW) is developed to compensate for errors and uncertainties in the forecast track. Because the MEOW represents maximum storm surge from several hundred storms, it compensates for any small forecast error in the track of a particular approaching storm. In the second phase, local NWS, state and local government, FEMA, and Army Corps of

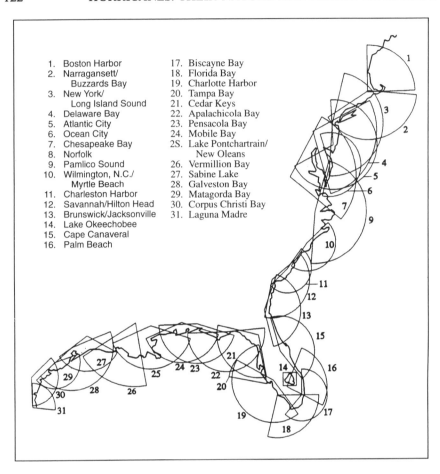

1. Boston Harbor
2. Narragansett/
   Buzzards Bay
3. New York/
   Long Island Sound
4. Delaware Bay
5. Atlantic City
6. Ocean City
7. Chesapeake Bay
8. Norfolk
9. Pamlico Sound
10. Wilmington, N.C./
    Myrtle Beach
11. Charleston Harbor
12. Savannah/Hilton Head
13. Brunswick/Jacksonville
14. Lake Okeechobee
15. Cape Canaveral
16. Palm Beach

17. Biscayne Bay
18. Florida Bay
19. Charlotte Harbor
20. Tampa Bay
21. Cedar Keys
22. Apalachicola Bay
23. Pensacola Bay
24. Mobile Bay
2S. Lake Pontchartrain/
    New Oleans
26. Vermillion Bay
27. Sabine Lake
28. Galveston Bay
29. Matagorda Bay
30. Corpus Christi Bay
31. Laguna Madre

Figure 5.2   The 31 SLOSH basins along the US Gulf and Atlantic coasts

Engineers officials attend a workshop held by the NHC on the use of the SLOSH model. The workshop covers development and application of the model. Finally, the US Army Corps of Engineers and various state and local officials complete evacuation studies based upon the SLOSH model estimates of inundated areas. The result of the application of the SLOSH model is a definition of the hurricane evacuation problem facing the particular SLOSH basin. Figure 5.3b illustrates the computed storm surge and observations associated with the landfall of Hurricane Hugo in 1989.

5.2.3 Winds

The strong winds of a hurricane can produce considerable structural damage and risk to life from flying debris, even inland from the coast. The damage

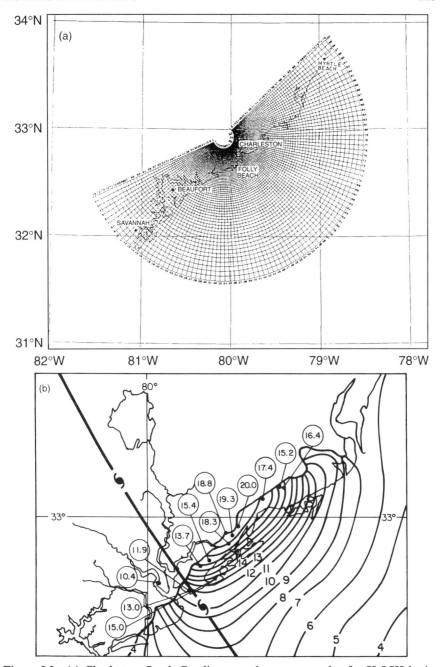

Figure 5.3   (a) Charleston, South Carolina as a close-up example of a SLOSH basin. (b) Computed storm surge (feet) of Hurricane Hugo (1989). Numbers in circles indicate observations or estimates of actual surge height. Note that the largest surge occurs to the right of the passage of the eye. Source: Jelesnianski (1993)

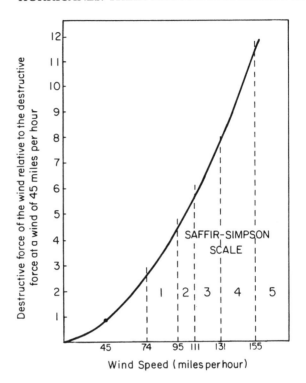

Figure 5.4   Relation between wind speed and its destructive force. For example, in the figure a wind of 135 mph (Category 4) has about nine times the destructive force of a wind 45 mph (tropical storm strength). In this plot, a destructive force of ϕ corresponds to a windspeed of 29 mph

caused by Hurricane Andrew was predominantly due to wind. Although winds reduce after landfall, as the central pressure of the storm lessens, destructive winds can still occur far inland.

The damage from winds is proportional to the energy of the airflow; thus, a wind of 100 mph is four times as effective at causing damage as a wind of 50 mph. This relation between wind speed and the destructive force of the wind is shown in Figure 5.4. Because of the destructive force of strong winds, intense hurricanes are responsible for most property damages related to hurricanes (see Section 5.3.1).

Maximum gusts, of course, are even stronger than reported sustained winds (which are measured in the United States by averaging wind speed over one minute). In a hurricane over the open ocean at about 36 feet (11 meters), a gust averaged over two seconds is generally about 25% greater than the one-minute average. For flat grassland, the two-second speed is around 35% larger, while in woods or cities, this measure of gust speeds is 65% greater. Thus a one-minute average wind of 100 mph would be expected to have gusts to 135 mph over the ocean and 165 mph over a forest. A wind of 100 mph over a forest at 36 feet, however, is harder to sustain than over the ocean at the same height, as a direct result of the greater gustiness over the forest. This larger turbulence acts to decelerate the wind as wind energy is transferred to the forest.*

5.2.4 Rainfall

Rainfall is often excessive at and after a tropical cyclone makes landfall, particularly if the very moist air of the storm is forced up and over mountains. These storms are particularly efficient rain producers when compared with other tropical and summer precipitation systems (Lawrence and Gedzelman 1996).

Rainfall from hurricanes is beneficial to agriculture, such as the rains from Hurricane Dolly (1995) in southern Texas and northeastern Mexico which relieved a drought (Rippey 1997, cf. Sugg 1967). Figure 5.5 illustrates observed rainfall associated with an August Atlantic tropical cyclone of 1928 as it moved up the Atlantic coast east of the Appalachians. Even relatively weak tropical-like disturbances can result in extreme rainfall, as seen, for example, over coastal Texas in September 1979 in which upwards of 19 inches (483 millimeters) of rain inundated the area over a period of several days (Bosart 1984). Occasionally, for reasons not completely understood, rainfall is light in the vicinity of hurricanes. Hurricane Inez in 1966, for instance, resulted in only a few drops of rain in Miami for several hours when the center was south and south-southwest of Miami and at its closest point to the city. At the time, Miami was under the wall cloud and, normally, torrential rains would have been expected. As a result of the absence of rain, the strong winds blew salt spray many miles inland, causing severe damage to vegetation from salt accumulation. Homestead Air Force Base, south of Miami and closer to the hurricane center, received only 0.62 inches (15.7 millimeters) of rain during the entire storm.

5.2.5 Tornadoes

Tornadoes are also a threat from tropical cyclones. Much of the damage of Andrew was associated with tornadic vortices whose wind speeds were added onto the large-scale hurricane winds (Black and Wakimoto 1994). These rapidly rotating small-scale vortices are spawned in squalls, usually in the front right quadrant of the storm with respect to the storm's track. While these tornadoes are not often as severe as the major tornadoes that are associated with springtime Great Plains thunderstorms, loss of life and property damage

---

* Powell, Houston and Reinhold (1996; Figure 4 of their paper) provide a framework to obtain standardized hurricane wind observations over land and over the ocean.

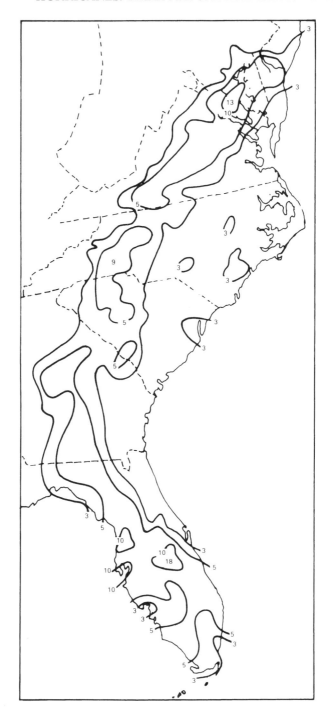

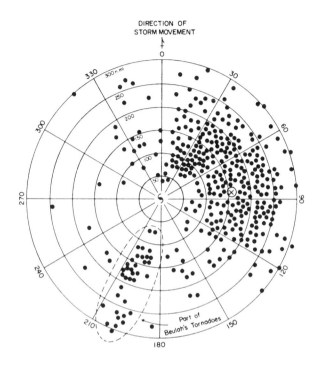

Figure 5.6   Observed locations of tornadoes with respect to landfalling storms. Source: Novlan and Gray (1974). Reproduced by permission of American Meteorological Society

does occur. Figure 5.6 shows observed locations of tornadoes with respect to the orientation of landfalling storms. Occasionally, landfalling storms produce tornado swarms, such as the tornadoes that were observed to be associated with the landfall of Hurricane Beulah (1969) on the Texas coast, as illustrated in Figure 5.7. Hurricanes Carla (1961) and Celia (1970), in contrast, produced no observed tornadoes.

5.2.6 Inland impacts

Inland, away from the coast, the largest threat to life and property occurs as a result of flash flooding and large-scale riverine flooding from excessive rainfall.

Figure 5.5   Observed rainfall in inches along the Atlantic coast of the United States from a northeastward moving tropical cyclone east of the Appalachian Mountains between 7 and 12 August 1928. Source: Dunn and Miller (1964). Reproduced by permission of Louisiana State Press

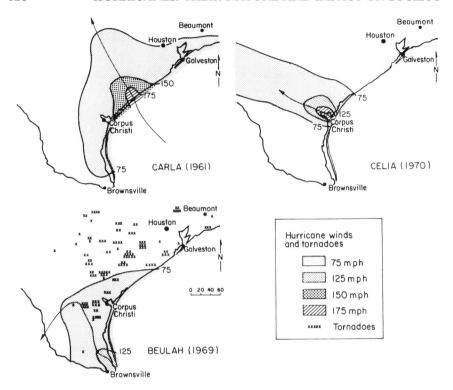

Figure 5.7   Tornado swarms associated with the landfall of Hurricane Beulah in 1969.
Source: Texas Coastal and Marine Council (1974)

Particularly dangerous are tropical cyclones whose rainfall is initially light and
benign after landfall only to erupt a couple of days later into torrential
downpours when the environment becomes favorable for the condensation
and precipitation of the large quantities of tropical moisture which have
moved inland with the storm.

A particularly extreme example of such a system is Hurricane Camille of
1969. After killing 139 people along the Gulf coast on 17 August, the storm
rapidly weakened after moving inland across Mississippi, into Tennessee and
Kentucky. There was relatively little concern by the National Weather Service
and certainly no hint of the tragedy that was to happen on the night of 19
August 1969 in central Virginia. The 24-hour and 12-hour precipitation forecasts for the area, for example, which are shown in Figure 5.8, indicate that
only slightly more than 2 inches (50 millimeters) were expected. Figure 5.9
shows the actual observed deluge that occurred in one part of Virginia as the
remnants of Camille began to rejuvenate through interaction with a cold front
and when the moist tropical air was lifted by the mountains. The rainfall of

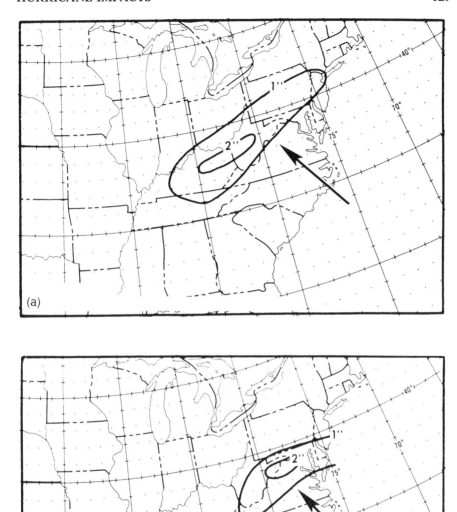

Figure 5.8   Forecast precipitation in inches ending 7 am EST 20 August 1969 for (a) 24 hours and (b) 12 hours. The location of the observed rainfall displayed in Figure 5.9 is indicated by the arrows

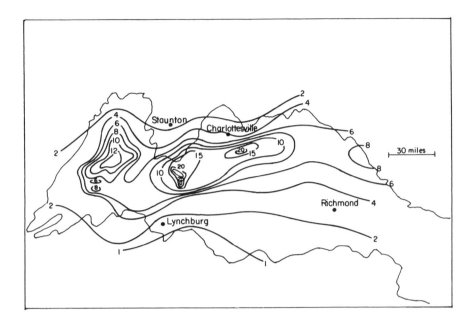

Figure 5.9   Observed rainfall in inches associated with the remnants of Hurricane Camille in central Virginia, 19–20 August 1969. The James River basin is outlined in the figure

almost 30 inches (760 millimeters) in six hours liquefied soils on the mountainous slopes and flooded drainage basins, burying and drowning 109 individuals. As a result of this tragedy, a radar site was installed in southern Virginia. One of the justifications of the new National Weather Service US doppler radar network (the WSR-D-88 system) is to detect such heavy rainfall events.

Such excessive rains well inland from landfalling tropical cyclones should be expected occasionally, as occurred in 1994 over Georgia associated with tropical storm Alberto. The environment of a storm is a localized region of the atmosphere which is enriched with water vapor, well in excess of even the average tropical environment. This occurs because the organized low-level convergence of moist lower tropospheric air into the hurricane over the ocean is transported upward, resulting in a deep layer of comparatively warm, near saturated air. After landfall, this rich reservoir of moisture moves inland.* This

---

* For example, as reported by Joe et al. (1995) from radiosonde soundings in the vicinity of Buffalo and Pittsburgh about one day after landfall of Hurricane Hugo in 1989, the moisture and temperature profile in the vicinity of the cyclone center retained its enriched tropical characteristics.

moisture can be copiously precipitated when it is lifted through a mechanism such as a mountain barrier and/or ascent over a weather front. Hurricane Agnes of 1972, for instance, produced enormous rainfalls over large areas of the middle Atlantic States because of strong large-scale atmospheric lifting and the movement of the moist air up and over the Appalachian mountains. The large-scale ascent, which was associated with a vigorous jet stream, would have produced an extratropical cyclone (but with more moderate rainfall) even in the absence of Agnes. The availability of the rich tropical moisture in combination with the lifting, produced disaster. The rainfall from this event is illustrated in Figure 5.10.

Even snowfall has been reported to be associated with the inland portion of a hurricane circulation. In 1963, Hurricane Ginny left more than 14 inches (36 centimeters) of snow in northern Maine as the hurricane moved into Nova Scotia with winds of around 100 mph (45 meters per second).

Wind damage and tornadoes also can occur well inland associated with tropical cyclones. In 1959, Hurricane Gracie caused 12 deaths in central Virginia 24 hours after landfall on the South Carolina coast. Hurricane Hugo in 1989 caused significant damage in Charlotte, North Carolina after landfall. From radar measurements, one study reported hurricane force winds around 3600 feet (1.1 kilometers) above the surface near Toronto associated with Hugo (although much weaker winds were felt near the surface) (Joe et al. 1995). Such strong winds aloft explain the substantial forest damage in the higher elevations in Shenandoah National Park in 1979 associated with the transit of the remnants of Hurricane David across the area, while no damage was reported in the lowlands.

5.3 SOCIETAL IMPACTS

5.3.1 Hurricane impacts on society

When they strike the US coast, hurricanes cost lives and dollars, and disrupt communities. Category 3, 4, and 5 storms – intense hurricanes – are responsible for a majority of hurricane-related damages. Appendix C shows intense hurricane tracks for landfalling storms over the US Atlantic and Gulf coasts for each decade this century. Loss of life, however, occurs from storms of various intensities. Owing largely to better warning systems, hurricane-related loss of life has decreased dramatically in the twentieth century (NRC 1989). Yet, in spite of reduced hurricane-related casualties ". . . large death toll in a US hurricane is still possible. The decreased death totals in recent years may be as much a result of lack of major hurricanes striking the most vulnerable

132

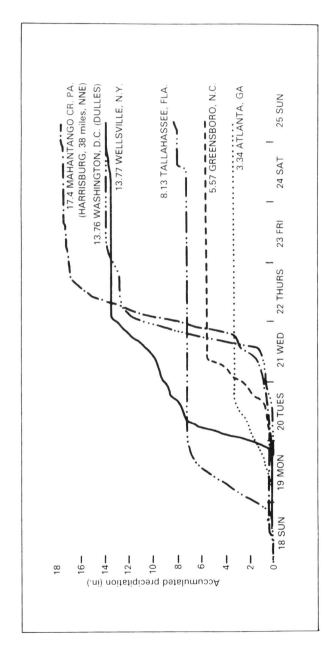

Figure 5.10   Cumulative rainfall in inches associated with Hurricane Agnes, 18–25 June 1972 for selected locations during the storm. Source: DeAngelis and Hodge (1972)

areas as they are of any fail-proof forecasting, warning, and observing systems" (Hebert, Jarrell and Mayfield 1993).

While loss of life has decreased, the economic and social costs of hurricanes are large and rising. A rough calculation shows that annual losses to hurricanes have been in the billions of dollars (cf. Sugg 1967; see Appendix B). In the United States alone, after adjusting for inflation, tropical cyclones were responsible for an annual average of $1.6 billion for the period 1950–1989, $2.2 billion over 1950–1995, and $6.2 billion over 1989–1995 (Hebert, Jarrell and Mayfield 1996). A recent study sought to "normalize" the damages associated with past storms to 1995 values; that is, to assess damages from past storms had they occurred in 1995 (Pielke and Landsea 1997). The study found $366 billion over 1925–1995, or about $5 billion annually (see Appendix B). For comparison, China suffered an average $1.3 billion (unadjusted) in damages related to typhoons over the period 1986–1994 (World Meteorological Organization, various years). Significant tropical cyclone damages are also experienced by other countries including those in southeast Asia (including Japan, China, and Korea), along the Indian Ocean (including Australia, Madagascar, and the southeast African coast), islands of the Caribbean, and in Central America (including Mexico). While a full accounting of global damages has yet to be documented and made accessible, it is surely in the tens of billions of dollars annually (e.g. Southern 1992).

Experts have estimated that worldwide, tropical cyclones result in approximately 12 000 to 23 000 deaths (Southern 1992; Smith 1992, Bryant 1991). Tropical cyclones have been responsible for a number of the largest losses of life due to a natural disaster. For instance, in April 1991, a cyclone made landfall in Bangladesh resulting in the loss of more than 140 000 lives and disrupting more than 10 million people (and leading to $2 billion in damages; Southern 1992). A similar storm resulted in the loss of more than 250 000 lives in November 1970. China, India, Thailand, and the Philippines have also seen loss of life in the thousands in recent years.

While the hurricane threat to the US Atlantic and Gulf coasts has been widely recognized, it has only been in recent years, following Hurricane Andrew, that many public and private decision-makers have sought to better understand the economic and social magnitude of the threat. Of notable concern is the vulnerability of industries with large production plants and facilities. For instance, according to R.H. Simpson, some years ago Dow Chemical shut down a plant as a hurricane approached, costing $10 million plus another $1.5 million for each day of lost production (R. Simpson 1997, personal communication). Also of concern is the storage of vast supplies of petrochemicals in vulnerable coastal locations on land very near sea level (R. Simpson 1997, personal communication). And finally, the insurance industry is also particularly vulnerable to hurricane impacts. (Appendix B provides data on hurricane damages and loss of life in the United States for the period 1900–1995.)

5.3.2 The challenge of estimating damages

In the aftermath of any extreme event there is a demand for a bottomline measure of damages in dollars. Yet there are many valid ways to measure the costs of a hurricane. Any assessment resulting in an estimate of total damages associated with a hurricane must be explicit as to the assumptions guiding the analysis of damages in order to facilitate interpretation of the estimate. The analyst needs to pay attention to at least five methodological factors, explained below, that can undermine attempts to construct a bottom-line assessment of damages: contingency, attribution, quantification, aggregation, and comparison.

Contingency: The problem of multiple-order impacts

A hurricane leaves in its path clear physical impacts on a community. (This section has benefited from the discussion by Changnon (1994).) As a result of storm surge, high winds, and rainfall, homes and businesses may be destroyed or damaged, public infrastructure may also be compromised, and people may suffer injuries or loss of life. Such obvious storm-related impacts can be called "direct impacts" because of the close connection between the observable damage and event. The costs associated with direct impacts are generally easiest to assess because they come in discrete quantities. Insurance payouts are one measure of direct storm impacts, as are federal aid, public infrastructure reconstruction, and debris removal (e.g. see Table 7.3).

Secondary impacts are those that are related to the direct impacts of a hurricane. Usually, secondary impacts occur in the days and weeks following a storm. For example, a hurricane may destroy a state prison. The direct impact is the cost of materials and labor associated with rebuilding the prison; secondary impacts might include the costs associated with housing prisoners elsewhere during rebuilding or delays in the legal system due to reduced prison space. Increases in insurance rates are another example of a secondary impact. In general, such secondary impacts are more difficult to assess because they require estimation based on assumptions and are also part of an existing social process; e.g. estimating the costs of delays in a legal process requires some sense of what would have occurred without the hurricane's impact.

Impacts on time-scales of months and years also occur. For example, a storm may destroy a number of homes in a community resulting in a decrease in property values, which in turn leads to a shortfall in property tax collection. As a result, community services that had been funded from property tax revenues may suffer, leading to further social disruption and thus additional costs (e.g. debris may not be collected for an extended period, and property values may thus decrease due to the appearance of the neighborhood). Estimation of the costs associated with such impacts is difficult to accomplish with much certainty because of numerous confounding factors.

In short, a hurricane is a shock to a community that leaves various impacts which reverberate through the social system for short and long periods. As the impact becomes further removed in time and in causation from the hurricane's direct impacts, pulling the signal of the reverberations from the noise of ongoing social processes becomes increasingly difficult.

Attribution: the problem of causation

Related to contingency is attribution. In the aftermath of a natural disaster people are quick to place blame on nature: "The *hurricane* caused billions of dollars in damages". However, it is often the case that the impacts of "natural" disasters are a consequence of human failures. For instance, damage is often a result of poor oversight of building practices and inadequate construction rather than simply the overwhelming forces of nature. Often, a disaster occurs at the intersection of extreme events and inadequate preparation. An important aspect of learning from the impacts of a hurricane is to understand what damages and casualties might have been preventable and which were not. Gross tabulations of damages neglect the question of why damage occurred, and often implicitly place blame on nature rather than ourselves, creating an obstacle to the drawing of accurate and useful lessons. An understanding of *why* damages occur is central to the drawing of useful lessons in the aftermath of a disaster.

Quantification: the problem of measurement

A hurricane impacts many aspects of society that are not explicitly or easily associated with an economic measure (e.g. psychological well-being). As a consequence, any comprehensive measurement of a hurricane's impact will necessarily include the quantification of costs associated with subjective losses. Therefore, the assumptions that one brings to an assessment of value can affect the bottom line. Care must be taken to make such assumptions explicit in the analysis. For instance, how much is a life worth? Or put in practical terms, how much public money are people willing to pay to save one more life in the face of an environmental hazard? According to a review by Fischer, Chestnut and Violette (1989), the public assigns between \$2.0 million and \$10.9 million as the amount of public funds it will devote to saving an additional human life (figures adjusted to 1993 dollars). The difficulties associated with assigning an economic value to a human life are representative of the more general problem of assessing many of the costs associated with a hurricane's impact that are not directly economic in nature. Similar questions might include: What is the value of a lost beach, park, or unrecoverable time in school, etc.? What are the costs associated with psychological trauma? If one seeks to quantify such costs (a difficult question in and of itself), then answers to issues like these require close attention to assumptions and methods underlying the analysis.

Table 5.1   Comparison of 1994 and 1995 hurricane seasons

	1994	1995
Named storms	7	19
Hurricanes	3	11
Intense hurricanes	0	5
Societal impacts		
Damage (billion US $)	>1.0	8–10
Deaths	1175	118

A comparison of the 1994 and 1995 hurricane seasons illustrates the problem of quantification. In terms of physical characteristics, the two years were very different: 1994 was one of the quietest years on record, while 1995 was among the most active. In terms of impacts, both saw extreme effects on society, yet in very different ways. In 1995, in the Atlantic Ocean basin, there were as much as $10 billion in economic damages due mainly to storms Luis, Marilyn, and Opal, while 1994 saw only $1 billion in losses. In 1994, however, 1175 people died, mainly in Haiti due to extreme flooding, while in 1995 118 died (see Table 5.1). The two seasons illustrate the different impacts that can be associated with different years: a quiet season does not necessarily mean less impact: consider that Andrew made landfall during the quietest four-year period in at least 50 years (Landsea et al. 1996).

Aggregation: the problem of benefits

Estimates of hurricane impacts usually concentrate on the costs and rarely consider the benefits associated with a storm (Chang 1984). Consider the following examples: the construction contractor called upon to repair and rebuild in the aftermath of a hurricane; an insurance company that is able to push through higher rates in the aftermath of a storm; a community able to restore itself to a level of productivity greater than it enjoyed prior to the storm with the assistance of federal disaster relief. Arguably, each realizes tangible benefits in some way from the storm. Should such benefits be subtracted from a storm's total cost? Another example is the sea wall that the city of Galveston built in the aftermath of the 1900 hurricane. In 1983 that sea wall saved lives and property from Alicia's unexpected fury (Chapter 1). Should the losses avoided in Galveston in 1983 as a direct result of the lessons learned and policies implemented in the aftermath of the 1900 disaster be attributed as a benefit of the 1900 storm? Such questions of benefits confound accurate determination of a bottom-line cost associated with a hurricane's impact. Assessments of damages are improved by specifying *who* suffers costs associated with a storm's impact. The net damage associated with a hurricane may be less important than the suffering of particular individuals and groups. Bottom-line assessments generally fail to consider who wins and who loses.

Comparison: the problem of demographic change

Communities that are vulnerable to hurricanes are undergoing constant change. People move. Homes and businesses are built. The storm that hit Miami in 1926 struck a very different place than the storm that hit Miami in 1992 (Andrew), even though, geographically, it is the same place. Thus, comparing hurricane impacts across time and space is problematic. Storms that make landfall in relatively sparsely populated areas would have certainly left a greater legacy of damages had they made landfall over a major metropolitan area. Yet, damage statistics often go into the historical record noting only storm name and economic damage (usually adjusted only for inflation). Such statistics can lead to mistaken conclusions about the significance of trends in hurricane damage. Because population and property at risk to hurricanes have increased dramatically this century, the statistics generally employed may grossly underestimate our vulnerability to hurricanes. Therefore, care must be taken in the use of bottom-line damage estimates as guides to policy conclusions based on a description of apparent trends. One method of "normalizing" past storm damages for today's conditions is given by Pielke and Landsea (1997).

The bottom line: apples with apples, oranges with oranges

There are many ways to measure the costs associated with a hurricane. There is no one "right" way. The method chosen for measurement of the cost of damages depends upon the purposes for which the measurement is made, and therefore must be determined on a case-by-case basis. No matter what method is employed when assessing the cost of damages associated with a hurricane, the analyst needs to ensure at least two things. First, the analyst needs to make explicit the assumptions which guide the assessment: what is being measured, how, and why. Second, compare "apples with apples and oranges with oranges." If the purpose is to compare the impacts of Camille with Andrew or even Hugo with Andrew, the analyst has to make certain that the methods employed result in conclusions which are meaningful in a comparative setting. The same caution ought to be observed when comparing Andrew with, say, the Midwest floods of 1993 or the 1994 Northridge earthquake.

5.3.3 Summary

To summarize, tropical cyclones impact ocean and land environments. In some cases, these impacts can be quite extreme in the form of wind and precipitation. Society's concerns about hurricane impacts result from economic damages, casualties, and other effects of storms. While intense hurricanes are historically responsible for most of the economic damages, large loss of life occurs due to weak hurricanes and even tropical storms and unnamed depressions. Assessment of hurricane impacts is important because it allows decision-makers to determine the relative significance of hurricanes in relation to the many other issues that vie for scarce resources. Assessment of actual and potential impacts sets the stage for society's response to tropical cyclones.

CHAPTER 6

Societal Responses

6.1 UNDERSTANDING SOCIETAL RESPONSES TO WEATHER EVENTS

In the 1970s, because of numerous weather-related impacts around the world, climate became of increasing interest to many decision-makers. Events that helped to stimulate this interest included the failed Peruvian anchovy harvest in 1971 and 1973, and the 1972–1974 drought in the African Sahel, a severe 1972 winter freeze in the Soviet Union, and in 1974 floods, drought, and early frost in the US midwest. In 1977, winter in the eastern US was the coldest ever recorded and summer was one of the three hottest in a century. As a consequence of these extreme events and their impacts, decision-makers began paying significant attention to the relation of weather, climate, and human affairs. Interestingly, Atlantic hurricanes were not one of the reasons for this increased interest, although the Indian Ocean basin did see extreme impacts during this period.

Understanding societal responses to weather and climate requires an understanding of the terms "weather" and "climate". The 1979 World Climate Conference adopted the following definitions of weather and climate:

> *Weather* is associated with the complete state of the atmosphere at a particular instant in time, and with the evolution of this state through the generation, growth and decay of individual disturbances.
> *Climate* is the synthesis of weather events over the whole of a period statistically long enough to establish its statistic ensemble properties (mean value, variances, probabilities of extreme events, etc.) and is largely independent of any instantaneous state.

Climate refers to more than "average weather" (Gibbs 1987). Climate is, in statistical terminology, the distribution of weather events and their component properties (e.g. rainfall) over some period of time, typically a few months to thousands of years. In general, climate statistics are determined based on actual (e.g. weather station) or proxy (e.g. ice cores) records of weather observations. Such a record of weather events can be used to create a frequency distribution that will have a central tendency, which can be expressed as an average, but it will also have a variance (i.e. spread around an average). Often, variability is just as or more important to decision-makers than the average (Katz and Brown 1992)!

What, then, is a "normal" (i.e. typical) weather event? There are different ways to define normal weather. In some cases, "normal" is implicitly defined as the climate over a recent 30-year period, e.g. as has been done with respect to floods (FIFMTF 1992). Of course, it is possible to argue that on planet Earth all weather events are in some sense normal; however, such a definition has little practical utility for decision-makers. One way to refine the concept is to define normal weather events as those events that occur within a certain range within a distribution, such as, for instance, all events that fall within one standard deviation of the mean (Pielke and Waage 1987). In practice, historical records of various lengths and reliabilities have been kept around the world for temperature, precipitation, storm events, etc. When data are available, such a statistical definition lends itself to equating normal weather with "expected" weather, where expectations are set according to the amount of the distribution defined as "normal". For example, about 68% of all events fall within one standard deviation of the mean of a bell-shaped distribution.

A change in the statistical distribution of a weather variable – such as that associated with a change in climate – is troubling because decision-makers may no longer expect that the future will resemble the past. For the insurance industry, as well as other decision-makers who rely on actuarial information, such a possibility of a changing climate is particularly troubling. A climate change is thus a variation or change in the shape or location (e.g. mean) of a distribution of discrete events (Katz 1993).

"Extreme" weather events can simply be defined as those "not normal", depending on how normal is chosen to be defined. For instance, if normal weather events are those which occur within two standard deviations of the mean, then about 5% of all events will be classified as extreme. While it is possible to define hurricanes as "normal" and "extreme", the simple fact is that for most communities any landfalling hurricane would qualify as an extreme event, because of their rarity at particular locations along the coast.

From the standpoint of those human activities sensitive to atmospheric processes, it is often the case that decisions are made and decision processes established based on some set of expectations about what future weather or climate will be like. Building codes, land use regulations, insurance rates, disaster contingency funds are each examples of decisions that are dependent upon an expectation of the frequency and magnitude of future normal and extreme events. Decision-making will take a different form in a situation where expectations are reliable and thought to be well understood than in a situation of relative ignorance (Camerer and Kunreuther 1989).

In short, many of society's decision processes are established based upon an expectation of "normal" weather. Yet for most coastal communities "normal weather" has historically (or at least over the time of a human memory) meant no hurricanes! Consequently, people are often surprised when a hurricane does strike and then overwhelms response capabilities. Because considerations of "extreme" weather are not always incorporated into regular processes of decision, when such events occur, they often reveal society's vulnerabilities and sometimes lead to human disasters. A fundamental challenge facing society is to incorporate weather and climate information into decision-making in order to take advantage of "normal" weather and to prepare for the "extreme". The degree to which society exploits normal weather and reduces its vulnerabilities to extreme weather is a function of how society organizes itself and its decision processes in the face of what is known about various typical and extreme weather events. The challenge is made more difficult by variability at all measurable time-scales in the underlying climate, and hence in the frequency, magnitude, and location of various weather events. And, of course, decisions that have a weather or climate component are also faced with all the political, practical, and social factors that influence policy.

6.2 LONG-TERM SOCIAL AND DECISION PROCESSES

In the constant battle between human action and nature's fury, society has developed a long track record of experience in mitigating and adapting to the impacts of extreme weather events.[1] Public and private hurricane policies in the United States have typically been successfully enacted in the aftermath of an extreme event that reveals societal vulnerabilities previously unseen or seen but not addressed. Long-term preparation involves actions in the face of the general hurricane threat facing a particular community, as opposed to preparations for a specific approaching storm. Generally, long-term preparedness involves (a) evacuation planning, (b) impact planning, and (c) recovery planning. In this discussion the term "decision-maker" refers to anyone faced with a decision in such processes. The term "policy-maker" is reserved for elected officials and administrators at various levels of government or business. Throughout, the term "policy" is used in a broad sense to refer to a "commitment to a course of action".

6.2.1 Preparing for evacuation

Evacuation involves moving people most at risk from a hurricane's impact out of the path of the storm. Successful evacuation depends upon plans put into place well before a particular storm threatens. Typically, the long-term component of evacuation planning has two parts: (1) a storm surge risk map, and

---

[1] The brief discussion of the various processes of hurricane preparedness in this chapter has benefited from the materials used in the hurricane preparedness course for local decision-makers organized by the FEMA Emergency Management Institute and the National Hurricane Center. The reader interested in learning more is encouraged to consult the sources in Appendix A.

(2) a technical data report. Educating the public about hurricane impacts and the evacuation process is also an important aspect of evacuation planning.

Technical data: behavioral, shelter, and transportation analyses

In addition to the storm surge maps, evacuation studies also include technical data reports on several factors relating to evacuation. The report generally includes a behavioral analysis, shelter analysis, and transportation analysis. The report indicates who is likely to evacuate, where they will go, and how they will get there.

The behavioral analysis is conducted using social science survey methods in order to assess the likely public response to various approaching storms. It provides information that is useful in estimating when people will evacuate, in what manner, where they would go, and what factors people consider important in deciding to evacuate.

The shelter analysis is conducted to ensure that there is sufficient capacity to house the portion of the evacuating population expected to seek refuge in a public facility. The shelter analysis tabulates existing public shelters, capacities, staffing requirements, public demand, and vulnerability to storm surge.

The transportation analysis is conducted to assess the amount of time necessary to evacuate people to safe locations in the event of an approaching storm. Planners develop evacuation zones or sectors for purposes of evacuation in order to determine and communicate the vulnerabilities to storm surge of particular residents and facilities, evacuation routes, and the location of shelters. The zones are devised based on the natural, demographic, and political features of a particular community and are made readily accessible to the public. Evacuation zones for Dade County, Florida, for example, are reprinted in the front pages of every phone book. It is necessary to develop zones that can be effectively communicated to the public in event of an evacuation. Major landmarks, roads, and easily identifiable natural features are often used to identify zone boundaries for the public, as zones are of little use if people do not easily know in which zone they reside.

The transportation analysis identifies the number of people and vehicles expected to evacuate from various zones under several scenarios, their likely routes, and roadway capacities and potential bottlenecks. The analysis also assesses anticipated time for evacuation prior to the arrival of tropical storm strength winds (>39 mph) and the likelihood of the inundation of roadways due to storm surge. It also explores the use of ferries, rail, air, and public transportation for purposes of evacuation.

6.2.2 Preparing for impacts

Coastal communities are at risk to hurricane-related effects of storm surge, wind, and rain. For example, communities generally develop building codes and land use regulations in preparation for these impacts of a hurricane. Of course, hurricanes also have environmental as well as societal impacts. For example, scholars have studied the geologic effects of hurricanes and the impacts of Atlantic hurricanes on south Florida mangrove communities (e.g. Coch 1994; Doyle and Girod 1997).

Building codes

Building codes are an important component in enabling structures to withstand certain forces of nature to which they are expected to be subjected to over their lifetime (or some defined period). Building codes, however, do not prepare for unexpected events or the most extreme. They provide a minimum level of protection, not a maximum level. In south Florida, Dade County developed the South Florida Building Code in 1957 to "address the need for hurricane-resistant construction" (FEMA/FIA 1993). Other structural factors are also addressed by the code, which has been long regarded as one of the toughest in the nation. It requires that structures withstand wind velocities of "not less than 120 mph at a height of 30 feet above the ground." Thus, winds in excess of 120 mph do not fall under the code's provisions. Figure 6.1 shows design wind speed for buildings and other structures along the US Atlantic and Gulf coasts.

Building codes can make a difference. According to John Mulady, an insurance industry official: "In the mid-1980s, two hurricanes of roughly equal size and intensity struck the coasts of Texas and North Carolina. A 1989 study of the damage done by hurricanes Alicia and Diana found nearly 70% of damage done to homes was the result of poor building code enforcement. However, in North Carolina, where codes were effectively enforced, only 3% of the homes suffered major structural damage" (Mulady 1994). As this study indicates, the mere existence of building codes is not sufficient for reduced vulnerability. Building code implementation, enforcement through inspection, and compliance are each essential factors to reduced vulnerability. It is important to note that the building code process occurs in a broader political environment of community policymaking and politics and that the code incorporates years of research experience in meteorology, engineering, etc.

Land use

It is widely accepted that "to mitigate many kinds of hurricane impacts government must be able to control certain uses of the land" (Salmon and Henningson 1987). Zoning, regulating, and taxing are common means of managing land use in areas vulnerable to hurricanes (Burby and Dalton 1994). Other methods include the creation of special property districts, in which, for example, a local government might exercise eminent domain, and public

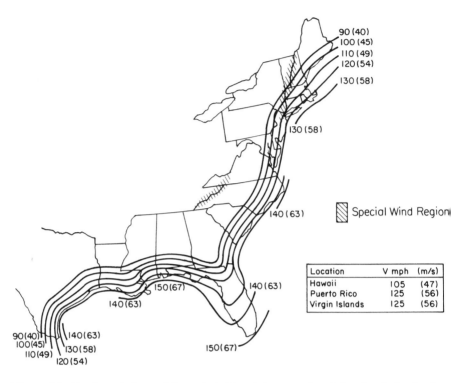

Figure 6.1 Design wind speed in mph (with m/s in parentheses) for the design loads for buildings and other structures along the Atlantic and Gulf of Mexico coasts (information provided by the American Society of Civil Engineers; figure provided by Leighton Cochran)

acquisition of land at risk. Interestingly, the land most at risk to hurricanes – that closest to the ocean – is also generally the most valuable from a standpoint of social desirability. Consequently, community efforts to buy land at risk can be thwarted by the high cost of beachfront property in demand by resorts and affluent individuals. Strategies of managing public and private land use are an important part of long-term efforts to reduce a community's vulnerability to hurricane impacts. The topic of land use is vast and much more complex than this simplification can present. For further reading, see Diamond and Noonan (1996) and the sources cited in Achter and McGowan (1984).

6.2.3 Preparing for recovery

Preparing for recovery involves a wide range of activities in anticipation of hurricane impacts. Two important components of recovery preparation over the long term are community planning and insurance.

Community planning

One study of hurricane recovery finds that "if a community 'returns to normal' without a [recovery] plan it is likely that many of the problems of the predisaster period will be recreated" (Salmon and Henningson 1987). A short list of activities that require planning over the long-term includes: building codes, flood control, beach conservation, and protection of telephone lines and water sources. Each of these aspects of a community's preparation for a hurricane's impacts can reduce its vulnerability and thus facilitate recovery efforts.

The study argues the case for "prior planning for post-hurricane reconstruction" (Salmon and Henningson 1987). In other words, communities threatened by hurricanes should assume that a hurricane will strike, and plan their recovery and restoration accordingly. They argue that "it is important that an outline listing the possible damage to the community and short-term, intermediate-term and long-term actions the government might consider be developed for review and consideration". It is important to note that "prior planning" for hurricane recovery and restoration generally occurs in an environment of "politics-as-usual" involving competing interests engaged in a political process of bargaining, negotiation, and compromise.

Insurance

Insurance is one technique that society uses to mitigate the economic losses associated with hurricanes (Roth 1996; Dlugolecki et al. 1996; BTFFDR 1995). Insurance works because each property owner who buys insurance pays a premium that in effect contributes to a collective fund that is available when a low-probability, high-impact event occurs. Because such extreme events impact only a very small proportion of insurance owners at any given time, sufficient funds are available to cover the losses. Based on their estimates of the probability of experiencing various losses, insurance companies set aside some portion of their funds to cover such losses. The remainder of the funds are used by the insurance company to pay for the costs of running its business and also to make money for its shareholders (RPI 1997). However, there may be instances when a catastrophic event like a hurricane results in more losses to a company than they have available to pay. Therefore, just as private property owners pay premiums to an insurance company to reduce their individual risk of catastrophic loss, insurance companies pay premiums to other companies to reduce their individual risk. This process is called reinsurance (Thorne 1984). In effect, reinsurance companies sell insurance to insurance companies; this process is repeated at different levels of risk in the global reinsurance industry. In this manner, risk of catastrophic loss is balanced across the world. "Thus, the total price of insurance should include the expected loss costs, the cost of doing business, the return on capital, and some profit" (RPI 1997).

HURRICANES: THEIR NATURE AND IMPACT ON SOCIETY

A central element in determining the expected cost of losses to hurricanes or any other peril is an insurance company's determination of risk (see Section 2.3.1 on risk assessment methodologies). Risk expresses an estimation of the probability of various levels of loss, and includes the uncertainty present in that estimate (Benktander and Berliner 1977). For example, as documented in Chapter 2, based on the historical record 1900–1994, in a given year, Dade County Florida has about 1 in 20 chance of experiencing a direct hit from an intense hurricane. That measure of 1 in 20 is of course not exact, so uncertainty must also be represented, perhaps by using a probability distribution. Of course, for actual rate setting, looking backwards to determine future hurricane incidence is fraught with methodological difficulties because of climate changes: hence, the insurance industry pays close attention to, and sometimes funds, the efforts of climate scientists in hope of better estimating the future. With an estimate of the future likelihood of a hurricane's impact, an insurance company would associate this probability with some estimate of expected dollar loss to their portfolio of insured properties. This process is typically based on a catastrophe model. Such models are often quite detailed in terms of climatology, engineering, and societal factors.

Risk, however, is not determined only in terms of the peril. In the United States, each state regulates the insurance industry in order to preserve company solvency and to ensure insurance availability and affordability (Roth 1996). Because of regulation, there is at times a political factor in rate setting which limits what insurance companies can charge and whom they can cover. For instance, a regulator's determination of a reasonable rate may be insufficient to cover the insurer's estimated losses. The goals of solvency and "availability and affordability" are thus at times at odds. This is one of the reasons why some companies have decided to restrict the sales of policies in coastal areas (e.g. Scism 1996). The reinsurance industry faces less regulation because it is typically located outside the United States.

The insurance industry's assessment of the risks that it faces due to tropical cyclones has changed dramatically in recent years. As recently as 1986 the insurance industry sponsored a study titled "Catastrophic Losses – How the Insurance System would Handle Two $7 Billion Hurricanes" which was presented as a "worst-case scenario" (Roth 1996). The study found that two such storms would have a serious impact on the industry. Only six years later Andrew struck south Florida resulting in about $16 billion in insured losses, with experts noting that much larger losses were narrowly avoided (Chapter 7). Primarily as a consequence of Andrew, the insurance industry began to reevaluate its exposure to hurricanes. Today, estimates of *insured* losses upwards of $50 billion are not uncommon (Chapter 2).

In the United States, property owners are insured by both the private insurance industry and the federal government. The private insurance industry is a for-profit enterprise, and focuses on insuring property owners against the effects of hurricane winds. The private insurance industry has traditionally not provided insurance for floods, including the effects of hurricane storm surge. Flood insurance has been the domain of the federal government through its National Flood Insurance Program (NFIP). After major flooding in 1951, Congress first became actively interested in the establishment of a Federal flood insurance program. After several years of false starts, major flooding in 1955 motivated passage of the Flood Insurance Act of 1956 (P.L. 84–1016). However, in part due to the opposition of the insurance industry, who felt that flood insurance, as a matter of principle, was impractical and actuarially unsound, the Act failed to win Congressional appropriations, and thus was never implemented (Mrazik and Appel Kinberg 1991). Congress did act following major flooding and flood losses associated with Hurricane Betsy in 1965. In 1966, the Task Force on Federal Flood Control Policy again recommended a Federal flood insurance program, but warned that "a flood insurance program is a tool that should be used expertly or not at all. Correctly applied, it could promote wise use of flood plains. Incorrectly applied, it could exacerbate the whole problem of flood losses" (TFFFCP 1966). Legislation was passed in 1968.

The National Flood Insurance Act of 1968 (P.L. 90–448) created the NFIP, which today is overseen by the Federal Emergency Management Agency (FEMA) through the Federal Insurance Administration. Congress mandated that the NFIP fulfil six primary objectives: (1) to make nationwide flood insurance available to all communities subject to periodic flooding; (2) to guide future development, where practical, away from flood-prone locations; (3) to encourage state and local governments to make appropriate land-use adjustments to restrict the development of land that is subject to flood damage; (4) to establish a cooperative program involving both the Federal Government and the private insurance industry; (5) to encourage lending institutions, as a matter of national policy, to assist in furthering the objectives of the program, and (6) to authorize continuing studies of flood hazards. By the end of the early 1990s, more than 2 million insurance policies were held in the approximately 180 000 communities that participated in the NFIP. Nevada is the only state with no communities in this program.

In theory, flood insurance is mandatory for property owners in flood-prone areas (Mrazik and Appel Kinberg 1991). However, as of 1993, in practice "just one in five mortgage holders in potential flood areas has federal flood insurance" (Benenson 1993). One reason for this was that almost half of the mortgages in the US were held by companies that were exempt from demanding flood insurance coverage on the mortgage. This loophole has been a source of criticism of the program. In addition, according to a 1995 Army Corps of Engineers' study, "while estimates vary it appears that 2 percent of [NFIP] policies have historically accounted for 25 to 50 percent of the dollars paid out from the National Flood Insurance Fund" (FPMA 1995, cf. BTFFDR 1995). Of course, while flooding associated with hurricane storm surge is only one of a number of types of floods that fall under the coverage of the NFIP, it does account for large losses.

6.3 SHORT-TERM DECISION PROCESSES

As a hurricane moves over the open ocean, scientists, elected and appointed public officials, news media, business owners, tourists, and coastal residents each begin processes of decision-making related to the approaching storm. Decisions that must be made include forecasts of a storm's likely future movement, where and when to post hurricane watches and warnings, whether to order evacuation, whether to respond to an evacuation order, etc. Ideally, the structure of short-term decision-making has been established through actions and preparations taken with a long-term view. Shortfalls in long-term preparedness efforts will become readily apparent in the face of an approaching storm as decision-makers face a shortage of time and other resources. Short-term decision processes are central to successful reduction of a community's vulnerability to hurricanes.

6.3.1 Forecast: the art and science of hurricane track prediction

A forecast of a hurricane's movement and future location is an essential component in the process of posting hurricane watches and warnings as well as in identifying locations to evacuate. The hurricane forecast is based upon the output of computer models and the subjective analyses of the forecasters (see Section 4.5). Robert Simpson, while director of the National Hurricane Center, observed in 1971 that:

> For decades hurricane forecasting has remained a product of subjective reasoning which varies with each forecaster's personal exposure to the hurricane problem, the "rules of thumb" he has developed, and the intuitive and analogue skills he has acquired.

Following Hurricane Andrew, the National Oceanic and Atmospheric Administration reminded forecast users that the forecaster's subjective judgment plays a determining role in the forecast process, in spite of the numerous technical and scientific advances of the past several decades. "Although NHC (National Hurricane Center) depends heavily on the use of [computer] model outputs, it is always the forecaster's judgment and experience that ultimately determines NHC's official track forecast" (DOC 1993). The forecast process remains, and will likely always be, both art and science (cf. Simpson 1978).

The NHC (one of three branches of the Tropical Prediction Center (TPC), which also includes the Tropical Analysis and Forecast Branch and the Technical Support Branch) routinely utilizes forecast models in the process of developing an official forecast track of an Atlantic hurricane for the public (DeMaria 1995; Sheets 1990; discussed in Section 4.5). The process of hurricane warning changes every year due to advances and refinements in the process. Thus the following discussion overviews the production and dissemination of forecast products as it existed in the early 1990s in order to provide a general sense of short-term hurricane response.

A number of models are used to forecast hurricane tracks because experience has shown that each has particular strengths and weaknesses that have been systematically documented and can be compensated for in the development of an official forecast. The forecaster's subjective judgment is necessary because "these models often provide conflicting information" (DeMaria, Lawrence and Kroll 1990). The result of the marriage of forecaster judgment and the various models' output is an "Official Forecast" for public dissemination.

As an example, Table 7.1 shows for Hurricane Andrew (1992) average forecast track errors for a number of models and the official forecast for various time horizons. Post-forecast assessments have judged the forecasts of Andrew's movement to be very accurate. For instance, the 24-hour average forecast error of 65 nautical miles was considerably less than the (1979–1988) contemporary 10-year average of 109 nm, and the 72-hour average error of 243 nm is significantly less than the 10-year average of 342 nm (DOC 1990). Note that in some instances individual models performed better that the official forecast, but that in general the official forecast out-performed the models.

The NHC forecasts hurricane intensity in addition to storm movement. Scientists analyze each storm in great detail as it approaches the US coast. They use data from satellites, reconnaissance aircraft, buoys, ship observations, radar, etc. to determine trends in storm motion, winds, and pressures. However, data such as surface wind observations are obtained infrequently because of the sparse information-gathering infrastructure in the vast Atlantic Ocean and Gulf of Mexico (Burpee et al. 1994). Data are gathered if a storm's path happens to pass over a National Data Buoy Center's moored buoy station, a Coastal-Marine Automated Network station, or a conveniently located ship (convenient for the forecaster, not the ship!). In past years, oil platforms in the Gulf have also hosted meteorological instruments (DOC 1993). The information gathered from such instruments is used in the production of forecasts of the storm intensification (Gray, Neumann and Tsui 1991; DeMaria and Kaplan 1994; DeMaria 1995).

To arrive at an official forecast of a hurricane's track and intensity, a number of officials participate in a conference call on the National Weather Service (NWS) Hurricane Hotline to arrive at the official forecast. Participating officials include: a National Meteorological Center (NMC) forecaster (expert on large models), forecasters at relevant NWS field offices (experts on local conditions), and the NHC hurricane specialist (expert on hurricanes). In addition, forecasters from the NMC Heavy Precipitation Branch and National Severe Storms Forecast Center (tornado expertise), officials from NWS National, Southern, and/or Eastern headquarters, and US Navy personnel may participate or listen in as appropriate. The tropical cyclone track and intensity forecasts from the different models are available for discussion. The final decision on the forecast results from discussion among these officials, although the NHC has final authority for the official forecast decision.

With the forecast in hand, discussion focuses on the potential impacts, timing, and locations of hurricane landfall. At this point, officials identify areas to post hurricane watches and warnings, which are issued by the NHC.

Using the forecast: hurricane watches and warnings

Hurricane watches and warnings are based on the official hurricane forecasts. A *hurricane watch* is an announcement for a specific coastal area that hurricane conditions pose a possible threat within 36 hours (Sheets 1990). A *hurricane warning* is an announcement that sustained winds of more than 74 mph associated with hurricane conditions are expected to affect a specific coastal area within 24 hours (Sheets 1990). A hurricane warning may remain in effect even if winds are less than 74 mph but a threat still exists due to high water or waves.

Upon landfall, hurricane force winds typically impact about a 50-mile wide swath of land. However, because of uncertainties in forecasts, hurricane warnings are generally posted for 125-mile segments of the coast: in a storm moving from east to west, the warning area would thus include about 50 miles south and 75 miles north of the expected point of landfall. The additional 25 miles to the north of the expected point of landfall are necessary because a storm is typically more intense to the right of center, with respect to its forward motion. A consequence of the need to post warnings for an area greater than the size of the storm is that people along about 75 miles of the coastline (or about 60%, on average) are warned of a hurricane but do not suffer the storm's greatest effects (i.e. 125 miles of warning area minus 50 miles of actual impact). This is called "overwarning" and is of concern because of the potential loss of public faith in forecasts and warnings and the practical problems of preparing a large community for the impacts of a hurricane.

The hurricane watch and warning process typically involves three groups: (1) NWS meteorologists, who bear statutory responsibility for watches and warnings; (2) relevant local and state officials who determine what response is needed; and (3) the media who are responsible for communicating watches and warnings to the public. Working together, the three groups try to satisfy three interrelated criteria. First, sufficient lead time must be provided for the community to protect life and property effectively. Second, care must be taken not to warn communities unnecessarily, which can potentially threaten the credibility of hurricane warnings – in other words, don't "cry wolf". Third, care must be taken to optimize response to the warning; that is, the warning should be timed and spatially distributed in such a manner as to evoke desired social actions and responses.

Watch and warning dissemination

Within an hour after decisions are made on actual hurricane warnings and watches, the NHC forecaster communicates this information to local and state officials and NWS officials via the National Warning System (NAWAS) hotline. State and local officials communicate back to the NHC any special information about their community that might affect the warning areas or timing. The final step is dissemination of the warning information to the public. The NHC disseminates three "products" which discuss the location, intensity, trends, and forecasts of tropical weather systems at various levels of technical detail (DOC 1993).

- *Tropical Cyclone Public Advisory (TCP)*
  The TCP is the least technical and most lengthy advisory. It is designed for use by the general public. It provides landfall probability forecasts for specific coastal locations.
- *Tropical Cyclone Marine Advisory (TMP)*
  The TMP is designed to provided technical and non-technical information to users in the marine and emergency management communities. It contains information on past and forecasted storm positions, predictions of wind strength, storm size, and current conditions.
- *Tropical Cyclone Discussion (TCD)*
  The TCD is designed to provide technical information to users such as the NWS, private consultants, the emergency management community, interests outside the US, etc.

The three types of advisories are disseminated every six hours, with TCP issued every two to three hours when landfall is imminent. A Tropical Cyclone Update (TCU) is issued periodically to disseminate information about changes in the storm system as conditions warrant. The NHC also issues special hurricane advisories and local NWS offices issue local advisories for particular counties. With rapid technological change, such as the development of the Internet, NHC products are likely to undergo significant changes in coming years. The Hurricane Center's web site (at http://www.nhc.noaa.gov) holds up-to-date information on these changes.

The chief conduit in the process of disseminating advisories to the general public is the media, and particularly television and radio. The local media is expected to play the role of a participant in the warning process, rather than that of an observer.

The actual evacuation

Convincing or coercing people to evacuate is often a difficult challenge that belies conventional wisdom. As one study has found: "it is clear that neither awareness of the existence of the hurricane hazard, nor indeed past experience with it, are sufficient to produce effective precautionary actions" (Baumann and Sims 1974). Furthermore, hurricane experience among coastal residents is rare. It was estimated in 1992 that approximately 85% of Gulf and Atlantic coastal residents have no experience with a direct hit from an intense hurricane (Jarrell, Hebert and Mayfield 1992). With continued coastal population growth, this proportion is likely to grow. It is probable that simple education of coastal residents regarding the serious effects of hurricane impacts will not be sufficient to ensure the evacuation of most people to safety. Moreover, laws concerning evacuations vary from state to state; uniform practices are thus probably undesirable and unattainable.

6.3.2 Impact: surviving the storm

In the final hours before a hurricane landfall, there is little that can be done to mitigate the storm's fury. Poorly built structures and people who do not evacuate will be severely tested by the storm's impact. NOAA (1993) offers "hurricane safety advice" for a hurricane impact.

- Store water:
  - Fill sterilized jugs and bottles with water for a two-week supply of drinking water.
  - Fill bathtub and large containers with water for sanitary purposes.
- Turn refrigerator to maximum cold and open only when necessary.
- Turn off utilities if told to do so by authorities.
- Turn off propane tanks.
- Unplug small appliances.

The guide advises people to "stay inside a well constructed building" and warns that "strong winds can produce deadly missiles and structural failure". If winds do become strong the guide advises:

- Stay away from windows and doors even if they are covered. Take refuge in a small interior room, closet, or hallway. Take a battery-powered radio, a NOAA Weather Radio, and a flashlight with you to your place of refuge.
- Close all interior doors. Secure and brace external door, particularly inward-opening double doors and garage doors.
- If you are in a two-story house, go to an interior first-floor room or basement, such as a bathroom, closet, or under the stairs.
- If you are in a multiple-story building and away from the water, go to the first or second floors and take refuge in the halls or other interior rooms away from windows. Interior stairwells and the areas around elevator shafts are generally the strongest part of a building.
- Lie on the floor under tables or other sturdy objects.

The guide warns that if the eye of the hurricane passes over, a calm period will ensue. However, the storm's fury will return in a matter of minutes with strong winds and heavy rain. When the eye passes over, people must realize that the storm's impact is only half over, and thus avoid being caught out in the open when the second half of the storm arrives.

6.3.3 Response: recovery and restoration

Community recovery and restoration in the aftermath of an extreme event with catastrophic social consequences have been studied extensively by scholars in the natural hazards research community (e.g., Haas, Kates and Bowden 1977). For a much richer and more contextual discussion of recovery and restoration following a disaster than can be given here, the reader is encouraged to consult the extensive literature from sociology, geography, and natural hazards on the topic (see, e.g. Drabek 1986 and May 1985).

Recovery

Community recovery from a hurricane event (as well as from disasters more generally) has two distinct phases. The first phase is the *emergency response period* in the immediate aftermath of a hurricane impact. During this period, which can last for a few days to a number of weeks, the "focus of action is on relieving immediate life-threatening conditions and restoring basic services" (Salmon and Henningson 1987). Successful emergency response depends upon the existence of a plan (developed well before the event) in order to avoid "ad hoc decision-making which could lead to unfortunate outcomes" (Salmon and Henningson 1987). In the aftermath of an extreme event such as an intense hurricane, the federal government may be called upon to assist in the recovery effort when local and state capabilities are exceeded.

From the federal government's perspective, key decisions that are made in the emergency response period following a disaster are focused around an assessment of needs and the provision of essential services. A determination of a community's needs, based on the extent of damage and the condition of disaster victims, is required to properly scale a response. The output from the assessment of needs is central in determining the ability of state, local, and volunteer organizations to respond. Victims' needs might include provision of food, water, shelter, medical attention, etc.[1]

The federal government's role in responding to a catastrophic natural disaster is triggered by a "Presidential Disaster Declaration", following a state

---

[1] Federal Emergency Management is the topic of a special issue of *Public Administration Review* (January 1985). More than 20 articles by leading experts in the field provide an accessible and valuable resource on the topic. More recently see BTFFDR (1995).

governor's request (Sylves 1996). The governor's request is based on a determination that adequate response to the disaster event is beyond the ability of the state. Following a disaster declaration, the Federal Emergency Management Agency (FEMA) is responsible for coordinating the activities of 26 federal agencies and the American Red Cross in response to the event. The American Red Cross, a private non-profit organization, acts as a pseudo-government agency in the context of disaster response. The role of FEMA and the other agencies is spelled out in a cooperative agreement between the agencies, called the "Federal Response Plan" (GAO 1993c). The Plan was extensively reworked following dissatisfaction with the federal response to Hurricane Hugo (1989). The modified version of the Response Plan was completed in April 1992. (See Section 7.4.1).

Restoration

A second phase in recovery from a disaster event is *restoration*, which can last from several weeks to a period of years. In the restoration phase a community has an opportunity to correct past mistakes that became evident during the disaster and to even take steps to improve the community beyond where it was prior to the disaster. During the restoration phase the context of decision-making changes: many more decisions must be made than under normal circumstances, putting time and other resource pressures on decision-makers. A decision whether to use "normal or extraordinary [community decision] procedures is itself a fundamental decision likely to create disruptive controversy" (Salmon and Henningson 1987). In addition, "groups or interests previously nonexistent (e.g., those with heavily-damaged property located in flood areas) may reshape community influence patterns" (Salmon and Henningson 1987). Thus, the restoration phase is one of opportunity and concern. It is a period of opportunity to correct past mistakes, but also a time of concern that past mistakes will be repeated or new ones made.

6.4 CONCLUSION: PREPAREDNESS ASSESSMENT

Short-term actions in the face of an approaching storm take place through processes of decision established and refined over the long term. Appendix E reprints a guide for local hurricane decision-makers prepared in 1992 by the Department of Emergency Management in Lee County, Florida. The guide lists 176 different actions related to hurricane forecast, impact, and response. The different actions – ranging from the straightforward to the complex – provide a useful starting point for assessment of a particular community's preparedness. Such an assessment might begin by simply asking whether the community is prepared to take each one of the 176 actions in the face of an approaching storm. If not, then long-term preparedness efforts might need improving. If a community is judged to have the capabilities to take each action, a next phase in the assessment might be to ask "How prepared are they?" In other words, the second and more difficult part of the assessment challenge would be to evaluate the health of the 176 different decision processes. Such an assessment has great potential to reveal unseen vulnerabilities to hurricanes, and engender communication and collaboration among groups that are generally isolated from one another.

CHAPTER 7

Hurricane Andrew: Forecast, Impact, Response

7.1 INTRODUCTION

As recounted in Chapter 1, Hurricane Andrew's impact in south Florida in August 1992 was by far the most costly hurricane in US history. Yet the worst case was avoided as Andrew made landfall in Dade County south of the area of greatest population and property. A review of south Florida's experience with Andrew provides useful guidance for improving preparedness in Dade County and other communities yet to feel the effects of an intense hurricane.

7.2 FORECAST

7.2.1 Hurricane track and intensity

Hurricane forecasters successfully predicted Andrew's track (Table 7.1). However, the storm's rapid speed and intensification were not anticipated. According to a report by the National Hurricane Center:

> ... on average, the NHC [official track forecast] errors were about 30% smaller than the current 10-year average. ... However, the rate of Andrew's westward acceleration over the southwestern Atlantic was greater than initially forecast. In addition, the NHC forecast a rate of strengthening that was less than what occurred during Andrew's period of rapid deepening. (Rappaport 1993)

The storm's rapid westward movement complicated preparation efforts by local emergency management officials. At 11:00 am, Saturday 22 August, the NHC forecast that tropical storm strength winds would arrive in Miami in 58 hours, or 9:00 pm Monday evening (DOC 1993). In the next NHC advisory, issued six hours later (at 5:00 pm Saturday), forecasters warned that tropical storm-strength winds would arrive in Miami in only 36 hours, or 5:00 am, Monday morning. Thus, in a six-hour time period emergency managers lost 22 hours of expected preparation time due to Andrew's unexpected rate of

Table 7.1   Hurricane Andrew average track forecast errors (miles) (after DOC 1993)

Model	1 Forecast period (hours)				
	12	24	36	48	72
Official	33	65	106	141	243
AVNO	60	75	89	97	132
BAMD	45	93	141	182	268
BAMM	40	81	121	151	229
BAMS	39	77	114	135	197
VBAR	32	60	93	138	287
CLIPER	35	81	148	233	437
NHC90	35	77	135	197	330
QLM	39	64	93	130	192
GFDL	36	71	93	117	209

forward speed. Because evacuation out of the region required 25 hours for a Category 2 or 3 storm and 37 hours for a Category 4 or 5 storm, the lost time made a decision to order evacuation imminent. The longer time is needed to evacuate for a more severe storm because more people are in the expected storm surge area.

Although hurricane forecasters expected a Category 3 storm at landfall, to be safe they cautioned emergency management officials to plan for a Category 4 landfall (DOC 1993). The margin of safety gained by preparing for one Saffir/Simpson category higher than is forecast is a general principle of hurricane preparedness. This margin of safety proved to be valuable as Andrew intensified more rapidly than was expected, and was, in fact, a strong Category 4 storm at landfall.

NHC forecasters were criticized by emergency management officials for the following announcement made on Friday 21 August at 5:00 pm: "Have a good weekend . . . tune back in on Sunday or Monday" (DOC 1993). That Friday afternoon the NHC hurricane track models suggested that then-Tropical Storm Andrew was slowing and indicated only a 7% chance of landfall in Miami by 2:00 pm Monday 24 August. As a consequence of the announcement, several emergency managers released staff for the weekend, staff that were later needed to implement evacuation orders that were issued Sunday morning.

The announcement and its unintended consequences are important for at least two reasons. First, the wording of the announcement and its interpretation by some as minimizing the threat from the storm highlight the difficulties of establishing effective communication between producers and users of forecast products. While producers of forecasts make every effort to present the information that decision-makers need for purposes of action, unintended messages are sometimes delivered. A second lesson is a reminder to users of forecasts that hurricane forecasting is both art and science. Hurricane forecasters have powerful and valuable tools and techniques at their disposal, yet knowledge of the future will always remain, to some degree, uncertain.

7.2.2 Evacuation

A FEMA assessment found that evacuation for Hurricane Andrew was successful, as judged by the relatively low loss of life due to the extreme event.

> The generally successful evacuation and short-term sheltering of south Florida residents during Hurricane Andrew can be attributed in part to pre-event planning and preparedness activities of the involved agencies. The Metro Dade County Emergency Operations Plan and the plans developed by local jurisdictions helped ensure their ability to respond in a timely manner. Considerable pre-event coordination and planning between the State, county, Army Corps of Engineers, the National Hurricane Center, and FEMA contributed to an understanding of the warning process, storm modeling, and requisite response actions. (FEMA 1993)

Although the evacuation for Hurricane Andrew was largely judged successful, there are a number of lessons to be learned from a number of less than optimal aspects of the evacuation process (USACE/FEMA 1993).

Evacuation orders were based on the hurricane warning that was issued at 8:00 am, Sunday 23 August, for portions of Florida's east coast. Dade County was ordered evacuated for a Category 3 storm at 8:15 am, about 19.5 hours prior to the passage of the eye of the storm over south Florida's Atlantic coast (DOC 1993). One hour later the evacuation order was upgraded to prepare for a Category 4 hurricane. The evacuation order was disseminated to the public by police driving through neighborhoods in the evacuation zone using loudspeakers as well as through the media. In addition to Dade County, St Lucie, Martin, Palm Beach, Broward, Monroe, Collier, Lee, and Charlotte counties issued evacuation warnings and approximately 750 000 people evacuated from south Florida (FEMA 1993).

Only Monroe County, immediately to the south of Dade County including the Florida Keys, issued evacuation orders prior to the NHC hurricane warning. Following the event, some emergency management officials in Dade County and in the Florida Department of Emergency Management criticized the NHC for waiting until 8:00 am to issue a hurricane warning rather than in its 5:00 am advisory (DOC 1993). "However, Dade County Emergency Management Director, Kate Hale, stated that her actions would have been no different had the warning been issued at 5 am or, for that matter, at 2 am" (DOC 1993). Thus, at least in Dade County, an additional three or six hours of warning time would have made no difference in the order to evacuate for

Andrew, largely because of the reluctance of emergency officials to order nighttime evacuations. It is important to remember that Andrew did not acquire hurricane-force strength until 48 hours before landfall in south Florida.

The officials who complained about the timing of the warning may have been concerned about their responsibility for a false alarm, had Andrew suddenly changed course, and wanted to base their decision on NHC information. In fact, in 1996, such a situation did occur with Hurricane Bertha. A hurricane warning was posted and the storm turned away from the coast. While Bertha was relatively well forecasted, the event illustrates the difficult trade-offs hurricane officials face. It is also possible that emergency officials may have had past experiences in mind. Hurricane Alicia (1983) strengthened unanticipatedly prior to landfall and an unprepared Galveston was saved from catastrophe only by its sea wall (Chapter 1). In September 1988, Hurricane Gilbert, the strongest hurricane of this century, was moving westward across the Gulf of Mexico. Emergency management officials in Galveston, Texas, ordered an evacuation based on a forecast issued by a private meteorological firm, even though the NHC official forecast had the hurricane passing well south of the Galveston area (Sheets 1990). The firm's forecast contained a significant error in its projected landfall point, placing the storm's track well north of where the models indicated it would track. By ordering a comprehensive evacuation, Galveston acted independently of its neighboring coastal and inland counties. The result was confusion and costs to the Texas coastal community. Dade County officials may have had the Galveston experiences in mind as they waited for the "official" hurricane warning before issuing evacuation orders.

In spite of these issues, according to the *Natural Disaster Survey Report* of Hurricane Andrew, produced by the US Department of Commerce, the NHC "watch and warning lead times during Hurricane Andrew were longer than average for landfalling hurricanes. That extra margin of safety was at least partially responsible for allowing hundreds of thousands of people to evacuate safely from south Florida" (DOC 1993). According to a study by E.J. Baker of Florida State University, 71% of residents in Dade County evacuated from high-risk areas. However, 28% of residents reported that they "didn't hear from officials that they were supposed to leave" (Baker 1993a). Official requests to evacuate, and the public's comprehension of that message, made a difference in decisions to evacuate. "Of those who indicated that officials told them to evacuate, 80% did, compared to only 52% of those who said they weren't told to leave" (Baker 1993a). Table 7.2 summarizes forecast and evacuation data from Hurricane Andrew (cf. USACE/FEMA 1993).

In addition to sufficient lead-time available for evacuation, the spatial distribution of the evacuation was also responsible for the relatively low casualty rate. However, evacuation of people out of the areas of greatest wind damage was largely a matter of good fortune rather than foresight.

Table 7.2   Summary of evacuation statistics for Dade County, Florida in Hurricane Andrew. Sources: Baker (1993a), DOC (1993)

Forecast and evacuation data	Dade County Evacuation Zones		
	Category 1	Category 2–3	Category 4–5
Evacuating population	227 210	408 740	589 000
Evacuation rate (% of total population)	0.71	0.63	0.33
Clearance time (hours)			
Within region	12	12	12
Out of region	15.5	25	37
Ideal watch time before landfall (hours)	NA	NA	36
Andrew watch	NA	NA	36
Ideal warning time before landfall (hours)	NA	NA	24
Andrew warning	NA	NA	21

Ironically, the areas that were evacuated in Dade County for expected Category 4 flooding, were devastated by the winds of Andrew – if the evacuation had not been carried out in those areas, the loss of life would have been much greater. (USACE/FEMA 1993, emphasis in original.) NA = Not Applicable

A consequence of Andrew's extensive wind damage was a finding that, due to the extensive impacts, Category 4 and 5 storms require a reassessment of vulnerability to high winds (USACE/FEMA 1993).

The SLOSH model process was judged successful, although human-made barriers not included in the model process changed the pattern of inundation from that which was predicted. Figure 7.1, reproduced courtesy of the *Miami Herald*, shows the storm surge associated with Hurricane Andrew in south

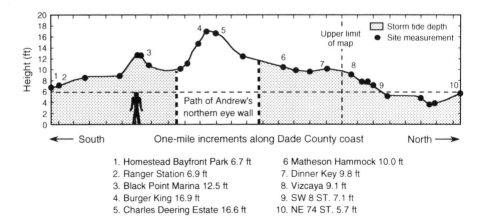

1. Homestead Bayfront Park 6.7 ft   6 Matheson Hammock 10.0 ft
2. Ranger Station 6.9 ft   7. Dinner Key 9.8 ft
3. Black Point Marina 12.5 ft   8. Vizcaya 9.1 ft
4. Burger King 16.9 ft   9. SW 8 ST. 7.1 ft
5. Charles Deering Estate 16.6 ft   10. NE 74 ST. 5.7 ft

Figure 7.1   Storm surge associated with Hurricane Andrew in south Florida. Source: redrawn from Doig (1992a). Reproduced with permission of *The Miami Herald*

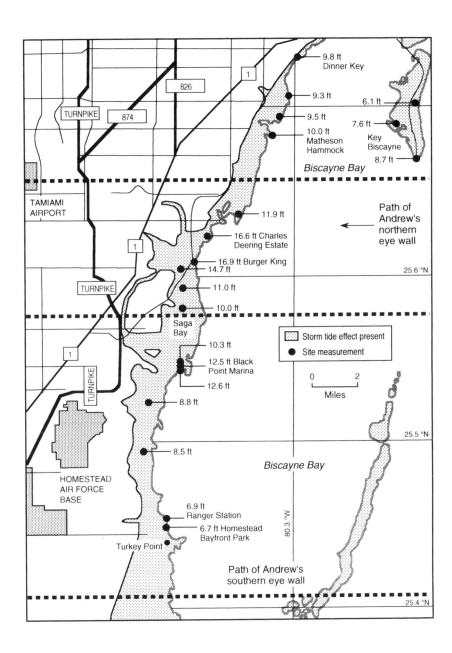

Figure 7.1 *(continued)*

Florida. A survey conducted following Andrew of the behavior of Dade County residents led to two conclusions about the accuracy of the behavioral analysis component of the Evacuation Technical Data Report.

• Hypothetical response data alone should not be used for driving evacuation behavioral assumptions; what people say they are going to do in a hypothetical scenario often diverges from their actual behavior, and
• A *variety* of assumptions should be provided for a variety of threat and evacuation scenarios, because public response will vary not only from one place to another in the same storm but from storm to storm in the same place. (USACE/FEMA 1993, emphasis in original)

The findings underscore "the fact that there is still much to learn about public response in hurricanes in general and the application of generalizations to specific locations" (USACE/FEMA 1993).

The shelter analyses led to the allocation of sufficient space for those seeking refuge in public shelters. In fact, throughout the warning area, many fewer people went to the shelters than was anticipated. With few minor exceptions, the shelter aspect of the technical data analysis was evaluated as successful (USACE/FEMA 1993). The transportation analysis also proved generally successful (USACE/FEMA 1993).

Evacuation decision-making prior to Andrew's impact was facilitated by the location of the NHC in Dade County. In the days preceding landfall, the Emergency Management Directors from both Monroe and Dade Counties (Billy Wagner and Kate Hale, respectively) interacted directly with NHC forecasters. In addition, the State of Florida Emergency Operations Center has a direct telephone link to the NHC. Outside of south Florida, NHC and emergency managers do not have the convenience of co-location, and communication may be more difficult.

One finding of the assessment conducted by the Department of Commerce was that:

Many coastal emergency managers do not understand the scientific reasoning involved in designating hurricane watch and warning areas. They want to evacuate either all or none of their coastal surge vulnerable area rather than parts of counties. (DOC 1993)

With such a large coastal population in south Florida, it is important to evacuate only those in designated areas. In many of south Florida's coastal counties, residents in different locations face different levels of vulnerability. Therefore, for some evacuation may be unnecessary, and should they evacuate, they might actually increase the challenge (e.g. due to gridlock on highways) of evacuating those facing a greater risk from the storm. As a result, the assessment recommended that the NHC work more closely with

emergency managers so that they understand better the watch and warning process.
One effect of the experience with Hurricane Andrew may be to exacerbate the evacuation problem facing south Florida. The assessment reported that

> As a result of increased anxiety caused by Hurricane Andrew, many south Florida residents indicated they would evacuate for future major hurricanes. Indeed, if this was the case, evacuation times for a Category 4 or 5 hurricane striking the Florida Keys would increase from the pre-Andrew level of 37 hours to 70–80 hours, depending on the percentage of residents evacuating. (DOC 1993)

This presents a problem because the NHC does not have the capability to reliably forecast landfall with a 70–80 hours lead time. In Table 7.1 the column of the 72-hour forecasts illustrates the large uncertainty in predictions of hurricane movement that far in advance. As residents become more knowledgeable about hurricanes, either through experience or education, the task of evacuation may become a more difficult policy challenge facing local communities, because too many people may try to evacuate, creating a gridlock situation where "Not everyone is going to get out in time. . . . Then we would see a real tragedy with many, many deaths. Not because people stayed in their homes when they should have evacuated. But because people left, when they should have stayed" (E.J. Baker, quoted in Van Natta 1992).
The task of educating the public about evacuation procedures is also made difficult by the public's relative lack of experience with hurricanes and the commensurate difficulty in estimating how people unfamiliar with hurricanes will respond to an evacuation order. Since hurricane activity has been lower than the historical average in the past several decades, if hurricane incidence increases evacuation rates may be difficult to estimate in future years (Baker 1993b). It is worth noting that any improved forecast capabilities will have little benefit if evacuation plans are not in place to capitalize on them. Studies of the evacuation problem are in need of constant update, as demographics, land use, and public perceptions undergo constant changes.

7.3 IMPACT

7.3.1 Direct damages from Hurricane Andrew

Estimates of direct damages related to Hurricane Andrew range from about $25 billion (e.g. Jarrell, Hebert and Mayfield 1992) to more than $40 billion (e.g. DOC 1993; Sheets 1994). Table 7.3 provides estimates of the various direct costs associated with Hurricane Andrew's impact in south Florida. The total damages directly associated with Andrew's impact in south Florida

Table 7.3   Breakdown of current dollar estimates of $30 billion in damages directly related to Hurricane Andrew in south Florida. Based on Rappaport (1993), updated and extended where possible

Type of loss	Amount ($ billions)	Source(s)/Notes
Common insured private property	16.5	Sheets (1994), includes homes, mobile homes, commercial and industrial properties and their contents, boats, autos, farm equipment and structures, etc.
Uninsured homes	0.35	*The Miami Herald*, 16 February 1993, reported in Rappaport (1993).
Federal Disaster Package	6.5	Anderson, Lim and Merzer (1992); represents 90% of $7.2 billion package (the rest went to Louisiana).
Public infrastructure		
State	0.050	Filkins (1994), tax revenue shortfall.
County	0.287	Rappaport (1993).
City	0.060	Tanfani (1992), Miami only.
Schools	1.0	Rappaport (1993).
Agriculture		
Damages	1.04	McNair (1992a,b)
Lost Sales	0.48	Fatsis (1992).
Environment	2.124	Rappaport (1993); includes state request for clean-up and repair of parks, marinas, beaches, and reefs.
Aircraft	0.02	Rappaport (1993)
Flood claims	0.096	FEMA Flood Insurance Administration, reported in Rappaport (1993).
Red Cross	0.070	Swenson (1993).
Defense Department	1.412	GAO (1993c) for DOD and US Army Corps of Engineers expenses during recover.

amount to about $30 billion. Although the tabulation is not comprehensive (e.g. city costs other than Miami's are not included), it is likely to be accurate to within 5%. Additional direct costs can certainly be identified and will increase this total, but it is the authors' judgment that such additional costs are likely to be less than $1.5 billion. If one were to add second-order and further costs, it is likely that total damages could approach $40 billion or even more, depending upon assumptions. As noted above, any tabulation of damages ought to note explicitly the assumptions guiding determination of further order impacts as well as how benefits are considered. Table 7.3 considers only direct costs associated with Andrew and ignores any benefits that might be related to the storm. Estimates of damages related to Andrew's impact in Louisiana are on the order of $1.0–1.5 billion. Andrew's impact on the national economy was the subject of some debate (see, e.g., Fields 1992).

With more than $16.5 billion in reported claims, insured losses represent the largest portion of the total. A number of property owners were uninsured (e.g. Swarns 1992). The estimated $1.04 billion in Andrew-related damage to the agricultural industry represents structural damage, lost crops and dead animals, and loss of the 1993 harvest (McNair 1992a). About 80% of Dade County's 3655 farms were damaged by Andrew (Anderson, Lim and Merzer 1992). At the time, only the 1988 drought in the Midwest had resulted in greater losses to agriculture ($3.4 billion). The storm "virtually wiped out" Florida's lime, avocado, tropical fruit, and nursery industries (Fatsis 1992). As bad as it was for farmers, it could have been much worse had Andrew struck later in the year, as farmers in Dade County had not yet begun planting the winter crop of tomatoes, squash, peppers, and beans (McNair 1992a).

Andrew completely destroyed about 63 000 of approximately 528 000 residences in Dade County (Finefrock 1992), and about 110 000 homes suffered some sort of damage (Anderson, Lim and Merzer 1992). Dade County property values decreased by approximately $3.0 billion in 1993, thus reducing property tax collections in 1993 by $50 million, which were made up for by the state (Filkins 1994). In 1994, however, Dade County property values increased by $2.9 billion, with "some properties coming back stronger than before," according to Dade County Manager Joaquin Avino (Filkins 1994). Within two years, Dade County property values had essentially returned to pre-Andrew levels. The one-year drop in property tax collection was more than made up for by an estimated $200 million in additional state sales tax revenue generated as a result of the large volume of building material sales during reconstruction (Silva and Nickens 1992).

Public and private infrastructure were severely damaged as well. Nine public schools were completely destroyed and 23 others were heavily damaged. University of Miami, Florida International University, and Miami-Dade Community College suffered about $60 million in storm-related damages (Anderson, Lim and Merzer 1992). According to one estimate, 1250 Dade County businesses suffered some type of damage, with small businesses of less than 10 employees in the majority (McNair 1992b). About 840 private tree nurseries also suffered heavy losses related to Andrew. Within two weeks of the storm 109 of 115 major hotels and resorts had returned to full operation (Fatsis 1992). Tourism emerged from the storm "relatively unscathed" with the Port of Miami, the airport, and destination resorts quickly reaching pre-storm operation levels (Fatsis 1992). In addition, Andrew caused environmental damage (USDI/NPS 1994).

According to Gary Kerney, Director of Property Claims Services, an insurance industry research group, the costs of Hurricane Andrew were exacerbated by the impact of Hurricane Hugo in South Carolina three years earlier (Noonan 1993). Hugo destroyed about 50% of South Carolina's pine lumber stock. This supply shortage, coupled with a depressed construction industry in Florida in the months preceding Andrew, set the stage for a shortage

Scenes of damage in Dade County in the aftermath of Hurricane Andrew. (Photos provided courtesy of C. Landsea)

of building supplies following Andrew. With higher demand for building materials, prices went up and rebuilding slowed. As a result of homes left in a damaged state during the rains and resultant water damage to structures that followed Andrew, "partial losses went from bad to worse, and many became total losses", further increasing the costs of Hurricane Andrew to the insurance industry (Noonan 1993). An interesting question is whether these costs ought to be attributed to Hugo or Andrew.

Hurricane Andrew provided some with economic benefits. Two weeks after the storm *The Miami Herald* announced that "The gold rush is on" as job seekers flooded south Florida (Alvarez 1992b). To some in the construction industry the storm came to be known as "St Andrew" due to the demand for construction that it left in its wake (Noonan 1993). The Miami city government expected a net financial gain due to the storm's second- and third-order impacts (Tanfani 1992). Clearly a number of public and private groups and organizations profited significantly from the storm's passage.

Although Andrew devastated South Dade County, "the hurricane's impact on Dade's economy could have been much worse," according to a report by the county's planning office (Filkins 1992). Andrew's path of most severe damage took it well south of the area that generates the most economic activity. Figure 7.2 shows how close Andrew came to being a much worse impact. The bulk of Dade County's businesses were not dramatically affected, which facilitated recovery. However, while the county planners expected a healthy recovery for Dade County in terms of aggregate economic indicators, they were much less sanguine about the future of southern Dade County, where Andrew had hit the hardest. They feared that communities such as Homestead would never fully recover.

Casualties

Difficulties in determination of a hurricane's economic impact are also present in attribution of casualties to a hurricane (cf. Garcia, Neal and Tanfani 1992). There are casualties directly associated with a hurricane, such as drowning due to a storm surge. There are also casualties that are indirectly related to a hurricane, such as a heart attack during the storm or being struck by lightning while sorting through damages in the days following the event.

The DOC Disaster Assessment attributed 14 deaths directly to Andrew's landfall in south Florida (DOC 1993). A FEMA assessment attributed 40 direct and indirect deaths to the storm (FEMA 1993). Hundreds of people were injured (DOC 1993). That there were not more casualties is partly due to good fortune, but also largely a result of long- and short-term preparedness efforts. The relatively low loss of life in Florida should not be interpreted to mean that future hurricanes do not have potential to take a large number of lives. As experts note, large loss of life is a constant threat from landfalling

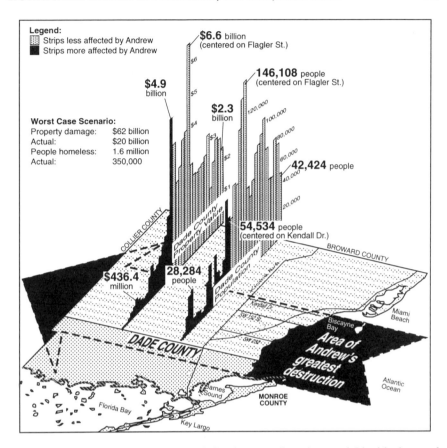

Figure 7.2   Property values and population in one-mile strips overlaid with the swath of south Florida most affected by Andrew. Source: Doig (1992b). Reprinted with permission of *The Miami Herald*

hurricanes, and coastal communities cannot let down their guard (AMS 1986; Hebert, Jarrell and Mayfield 1993).

In addition to physical injuries, victims also experienced psychological trauma (Naunton 1992). According to Dr Charles Gibbs, a clinical psychologist and chairman of the Dade County Crisis Response Task Force, "The survivors are suffering a myriad of problems". For instance, he noted that:

> There's a lot of anxiety, restlessness, and an increase in anger and irritability – all of which is a perfectly normal response to a disaster of this magnitude. Many of the children are suffering night terrors. They see "Andrew" as a real person and they're afraid he's coming back, that he's going to kill them. People feel like they're in a war zone down here – there are 18 000 troops, and the air is filled with helicopters. (Quoted in Mauro 1992)

While the psychological effects of Hurricane Andrew were especially hard on children, adults also experienced depression and Post-Traumatic Stress Syndrome, which can lead to sleep disorders, intense guilt, and phobic reactions (Mauro 1992).

7.3.2 Building codes: construction, implementation, enforcement, and compliance

Because Dade County's building codes have long been considered by many to be among the strictest in the nation, much of the widespread damage following Andrew was unexpected. Investigations into the causes of the damage revealed that not only was it unexpected, but unnecessary as well.

Unexpected patterns of damage

The unexpected extent of wind damage led to some of the most important lessons of Hurricane Andrew in south Florida. According to one assessment:

> The South Florida Building Code is one of the toughest in the country, and all observers agree that if Hurricane Andrew had hit almost any other location on the U.S. coastline, deaths and damage would have been much higher due to a greater number of building structure failures. (Levy and Toulman 1993)

Yet, even though south Florida does have a strong building code, Andrew revealed extensive flaws in the code implementation and enforcement process, as well as in the code itself. According to insurance industry estimates, poor compliance with the south Florida building codes accounted for 25% to 40%, or about $4 billion to $6.5 billion, of the insured losses in south Florida due to Hurricane Andrew (Noonan 1993).

That Dade County has long been considered to have among the most rigorous and well-enforced hurricane-related building codes should give pause to the other 167 coastal counties from Texas to Maine. If Dade County is indeed the best prepared, and it suffered at least $4 billion in preventable damages, then there are almost certainly great opportunities to reduce the potential for property damage elsewhere along the Atlantic and Gulf coasts. However, even with strict enforcement of the South Florida Building Code, Andrew would still have been the costliest hurricane in history. Thus, it is indeed frightening to consider the potential impacts of an Andrew-like storm on other populous coastal communities along the US Gulf and Atlantic coasts.

According to a joint Federal Emergency Management Agency/Federal Insurance Administration assessment of building performance in Hurricane Andrew, excessive damage occurred due to a number of factors.

The breaching of the building envelope by failure of openings (e.g., doors, windows) due to debris impact was a significant factor in the damage to many buildings. This allowed an uncontrolled buildup of internal air pressure that resulted in further deterioration of the building's integrity. Failure of manufactured homes and other metal-clad buildings generated significant debris. Numerous accessory structures, such as light metal porch and pool enclosures, carports, and sheds, were destroyed by the wind and further added to the debris. (FEMA/FIA 1992)

In addition to structural factors in residential building, the assessment placed responsibility for the building failures on breakdowns in the broader processes of residential construction and building code compliance and enforcement.

Much of the damage to residential structures resulted from inadequate design, substandard workmanship, and/or misapplication of various building materials. . . . Where high quality workmanship was observed, the performance of buildings was significantly improved. Inadequate county review of construction permit documents, county organizational deficiencies such as a shortage of inspectors and inspection supervisors, and the inadequate training of inspectors and supervisors are factors that may have contributed to the poor-quality construction observed. (FEMA/FIA 1992)

In short, the excessive damage seen in south Florida resulted from a broad breakdown in the process of code enforcement and compliance.

An investigation by *The Miami Herald* revealed in detail the breadth and depth of the failures in the building code decision process. The *Herald* conducted a three-month investigation during which they analyzed 60 000 damage inspection reports for 420 south Florida neighborhoods. When they superimposed their neighborhood damage map over a wind speed map, "like a latent fingerprint found at a crime scene, a clear pattern has appeared in the vast sprawl of destruction left by Hurricane Andrew" (Leen et al. 1992). Figure 7.3 shows a much simplified map of destruction superimposed on a simplified wind speed map. The *Herald* investigators found that

- Many of the worst hit neighborhoods were far from the area of maximum winds.
- Some nearly destroyed neighborhoods were adjacent to neighborhoods with relatively light damage.
- Newer homes, built since 1980, suffered damage at a greater rate than did older homes.

The *Herald* investigation blamed the pattern of unexpected destruction on a broad "breakdown" in the building code enforcement process (Getter 1992b). The failure manifested itself in poor construction, that is, the construction of buildings that did not meet the intent of the South Florida Building Code. As

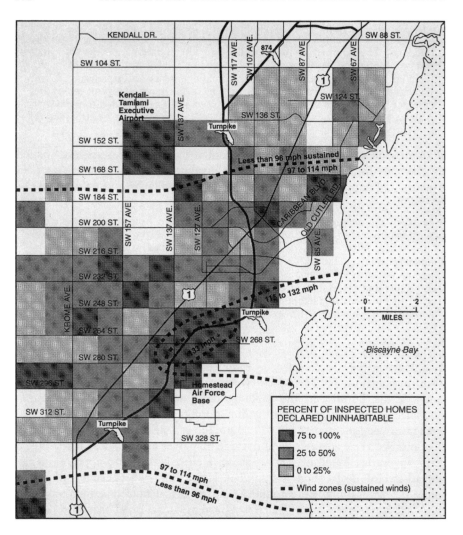

Figure 7.3   Map of destruction superimposed on a wind speed map. Source: *Miami Herald* Staff (1992b). Reprinted with permission of *The Miami Herald*

a consequence, Andrew's damage in Dade County was worse than it might have been had the provisions of the code been met. The *Herald* investigation attributed the poor construction to a period of 15–20 years during which complacency about hurricanes led to less rigorous enforcement of the building code, as well as an actual weakening in the code (Leen et al. 1992).

The *Herald*'s investigation placed the responsibility for the "failure of design and discipline" in part on the political process that allowed builders in Dade County to have "a considerable influence in the department that

POLITICS OF THE BUILDING CODE

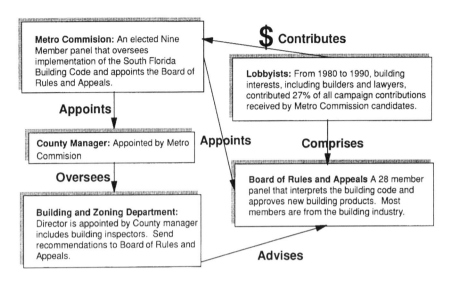

Metro Commision: An elected Nine Member panel that oversees implementation of the South Florida Building Code and appoints the Board of Rules and Appeals.

Appoints

$ Contributes

Lobbyists: From 1980 to 1990, building interests, including builders and lawyers, contributed 27% of all campaign contributions received by Metro Commission candidates.

County Manager: Appointed by Metro Commision

Appoints

Comprises

Oversees

Building and Zoning Department: Director is appointed by County manager includes building inspectors. Send recommendations to Board of Rules and Appeals.

Board of Rules and Appeals A 28 member panel that interprets the building code and approves new building products. Most members are from the building industry.

Advises

Figure 7.4   Relationships between various participants in the South Florida Building Code Decision process. A *Miami Herald* analysis found a conflict of interest. Based on Getter (1992b)

inspected them" (Figure 7.4). At the height of the building boom, the building industry contributed "one of every three dollars" to campaigns for the Metro Commission that oversees and interprets the South Florida Building Code (Leen et al. 1992). The Metro Commission appoints the members of the Board of Rules and Appeals, which has responsibility for determining the suitability of new construction materials. In the years leading to Andrew, most of the Board's members came from the building industry (Getter 1992c).

The findings of the *Herald*'s investigation can be summarized in terms of complacency, accountability, and communication.

Complacency   Complacency is reflected in the lack of attention paid by building inspectors and government officials to actual inspections. For instance, in 1988 Dade County employed 16 building inspectors to serve a population of well over one million. On many occasions in the years preceding Andrew, inspectors reported conducting more than 70 inspections per day, a rate of one every six minutes, not counting driving time (Getter 1993). Some did not even leave their cars to conduct inspections. One hurricane expert, Peter Black, aptly summed up the results of poor construction that went undetected: "The damage is directly proportional to the kind of construction used" (quoted in Leen et al. 1992). It is likely that complacency was a consequence of several decades of low hurricane incidence in south Florida. As one veteran engineer commented, part of the problem was that "people were just oblivious to things, as if they thought we never were going to have a hurricane in this area" (quoted in Leen et al. 1992).

Complacency was pervasive, and included home buyers as well as home builders. As an engineering professor at the University of Miami noted: "inspection is the second line of defense in this industry, you may want to blame it all you want, but it's supposed to be built right in the first place" (quoted in Leen et al. 1992). The failure in the Dade County building code process is attributable to complacency throughout the system – including the public, their elected representatives, appointed officials, and private industry.

Accountability   Complacency persisted due to a lack of accountability. In addition to building inspectors, the builders, government officials, and home buyers also share responsibility for failing to ensure that building codes were properly implemented. The magnitude of Andrew's destruction should not come as a surprise to anyone in the south Florida community. Grand jury investigations called in 1976, 1986, and 1990 to assess the state of building code implementation, revealed many flaws in the process. The 1976 grand jury suggested that "inspection practices of the last several years have resulted in construction of buildings which could be blown away in another '1926 Hurricane.' . . . We are concerned about such a situation" (*Miami Herald* Staff 1992a). A 1990 grand jury found that "our investigation into construction problems and building inspections indicates that stricter controls are needed to assure the continued integrity of construction" (*Miami Herald* Staff 1992a). In addition, the following conclusions were drawn in the aftermath of the 1926 Miami hurricane:

> Well-built structures weathered the hurricane except for minor damage such as broken windows and displaced roofing and shingles, while those of substandard construction were generally wrecked. . . . The older frame houses generally withstood the storm, but many of the small frame buildings, erected during the boom period, with little regard for proper bracing, were blown to bits early in the storm. (Quoted in Cochran and Levitan 1994)

Communication   One important part of the challenge of securing accountability and thus effectively implementing building codes arises from a lack of communication between different groups in the decision process. For instance, in 1992 the federal building code for mobile homes, which prohibits state or local officials from establishing their own standards, did not mention wind speeds, and instead set its standards on a structure's ability to withstand pressure. As a consequence of a lack of agreement on how to measure a structure's ability to withstand pressure, "the mobile home industry and critics differ on how to translate pressure into wind speeds" (Whoriskey 1992). Federal, state, and industry officials had long argued that the federal code translated to protection against 110 mph winds. A 1989 Texas Tech study found the code only protects to 80 mph. A NIST study of the damage related to Hurricane Andrew supported the findings of the 1989 study (Whoriskey

1992). Since Andrew, the code has been changed. Another example is provided by the differences in units of wind speed measurement used for building
code and meteorological purposes. The NWS uses the terms "sustained wind"
and "highest gust" while building codes often use the term "fastest mile". A
sustained wind is the average over a specified time period. The NWS uses a
one-minute average. Recently, the three-second gust speed has been adopted
for building design. While conversion between the two units is possible, "the
issue is confusing to the general public" (FEMA/FIA 1992). A central part of
improving implementation of building codes is to ensure that effective
communication exists between all parties that interact in the building code
process. If communication is successful, then failures can be better traced to
purposeful evasion of the law.

   Builders: "It was Nature's Fault"   In general, builders denied culpability
following the storm, and placed blame on "a hurricane that exceeded the
code" (Leen et al. 1992). The South Florida Building Code mandates that
buildings resist sustained 120 mph winds. Thus, extensive damage that
occurred in areas of greater than 120 mph winds would not be due to
problems with the building code process. One building contractor noted that:
"When you're hit with winds of up to 160 miles per hour, something's got to
give". The Mayor of Miami upped the ante: "There's nothing we could build
today with home construction that would stand. I suspect that the wind was
over 200 miles per hour" (Leen et al. 1992). Comments like these spurred a
debate over how fast Andrew's winds actually were. Determination of wind
speed would provide clear evidence as to whether the excessive damage related
to Andrew was due to human failures or nature's fury.

   Determination of wind speed, however, was made difficult because of the
lack of available data. The NHC lost its radar and many instruments failed
during the storm. However, subsequent analyses pieced together individual
anemometer readings, available radar imagery, and damage patterns to
determine wind speeds (e.g. Powell, Houston and Rheinhold 1996; Powell and
Houston 1996; Wakimoto and Black 1994; Rappaport 1993; Fujita 1992).
Scientists determined that a small region near the center of the storm experienced sustained winds (averaged over one minute) of 145 mph, with gusts (over
two seconds) to 175 mph, corresponding to a Category 4 hurricane. Beyond the
relatively small area of greater than 120 mph winds, most of Dade County
experienced sustained wind speeds of less than 120 mph. Thus, Andrew was not
a force of nature beyond experience. For the most part, the storm's wind
impacts were within the limits prescribed by the South Florida Building Code.
Excessive damage was in many cases the result of human failures.

   Preventable suffering   In short, a significant degree of property damage
related to Hurricane Andrew was the result of a breakdown in the process of
building code implementation and enforcement. A Dade County grand jury,
convened in 1992 to explain the unexpected damages related to Andrew,
concluded that:

While we, as a community, have suffered greatly as an unavoidable result of Hurricane Andrew, this suffering was aggravated by the systematic failure of our construction industry and building regulation process. Had the failures not existed much of the suffering would have been prevented. (Quoted in Leen et al. 1992)

In his testimony before Congress on 2 February, 1994, Robert Sheets, director of the National Hurricane Center, summarized the lessons of Hurricane Andrew for building practices in coastal communities beyond Dade County.

People thought that the South East Florida Code was adequate. However, some small changes in the code and interpretation of the code had crept in which caused massive losses and some loss of life. Even with these deficiencies, that code and building practices in Dade County were far superior to most other hurricane prone areas. It often is a problem of education rather than the small increase in cost required to have a good code in place. . . . It seems that an adequate experience level exists today, at least from a technical standpoint, where an effective, relatively low cost, uniform code, could be adopted and applied for each type of structure along the coast. . . . However, such codes and enforcement practices are rare in most coastal areas. (Sheets 1994)

7.3.3 Insurance

Hurricane Andrew served as a "wake-up" call to the insurance industry. Of the 371 insurers who reported Andrew-related losses, nine declared insolvency (Changnon et al. 1996). However, in spite of Andrew's record impact on insured property, in aggregate, the worldwide industry remained profitable in 1992 (BTFFDR 1995). This is the same year as Hurricane Iniki in Hawaii ($1.6 billion in insured damages), tornadoes and hail storms ($3 billion), ice storms ($1 billion), the Los Angeles riots ($775 million), and a Chicago flood ($300 million). Andrew's impact on the insurance industry was to call attention to what easily could have happened had it hit an area of greater concentration of insured property.

As a consequence of Andrew, insurers immediately attempted to limit coverage in areas subject to hurricane impacts and raise rates where coverage was provided (Roth 1996). A number of insurers completely stopped granting insurance coverage in Florida. In response, the Florida State Legislature took steps to limit restrictions on the granting of insurance (BTFFDR 1995). The State has also established an insurance fund to insure those who would otherwise go without (ATFHCI 1995).

The difficulties that the insurance industry faces in raising its rates to better match its exposure to hurricanes are illustrative of the challenge of changing business-as-usual, even in the aftermath of an extreme event like Andrew. The executive vice-president of a large business-insurance company wrote a memo on the day of Andrew's impact in south Florida that identified the event with an opportunity to raise insurance rates based on his assessment that "The industry cannot absorb the [Andrew] loss and the cash hit without increasing rates" (quoted in Garcia and Satterfield 1992). The public and media responded to the leaked memo angrily. Reaction to the memo reflects fundamental difficulties in insuring property in areas subject to risk: Among whom should the risk be spread? For instance, should people in Kansas pay insurance premiums that reflect the costs of rebuilding oceanfront homes? Should insurance rates for coastal residents be scaled to the expected annual losses due to hurricanes? If so, rates would likely be higher then they have been in the past. Should people be allowed to live in vulnerable areas without insurance? These are difficult questions that have no simple answers, but require attention in the process of improving society's hurricane preparedness.

7.4 RESPONSE

7.4.1 Recovery

A test of federal, state, and local emergency management

Hurricane Andrew was the first time that the federal government's "Federal Response Plan" had been used since it had been extensively modified following Hurricane Hugo (1989). Beginning with Dade County Emergency Manager Kate Hale's cry for help, broadcast in giant type on the front page of *The Miami Herald* several days after impact (Chapter 1), the federal government's initial response to Andrew was widely criticized. A General Accounting Office evaluation of the implementation of the Federal Response Plan following Hurricane Andrew's impact found a number of flaws.

> The plan lacks, among other things, provisions for a comprehensive assessment of damages and the corresponding needs of disaster victims. In addition, the response in South Florida suffered from miscommunication and confusion of roles and responsibilities at all levels of government – which slowed the delivery of services to disaster victims. (GAO 1993a)

The official assessments of the federal response were aptly summarized by a frustrated Dade County official who characterized the federal response as "a whole lot of resources but no coordination" (Viglucci 1992).

One lesson in the aftermath of Andrew was a need for better communication and coordination between meteorologists and agencies responsible for federal disaster response, particularly FEMA. Thirty hours elapsed between the issuance of a hurricane warning for south Florida and the declaration by President Bush of a major disaster, nine hours after Andrew made landfall (FEMA 1993). While response activities did begin before the formal disaster declaration, better communication and coordination between the forecasters and emergency management officials may have facilitated recovery efforts in the immediate aftermath of the event. A number of assessments of the Andrew experience recommended ways to prepare for some type of declaration of disasters in advance in order to better mobilize federal support (e.g. FEMA 1993; DOC 1993; Levy and Toulman 1993). The Chief of Planning for the Florida Department of Community Affairs stated in Congressional testimony that "we need a greater ability and flexibility to not only identify Federal assets, but mobilize these assets prior to a hurricane making landfall" (Koutnik 1993). Hours before Andrew struck, information was available that indicated that a disaster was imminent. However, because of uncertainty as to whether state and local expenditures to prepare for a pending disaster will be reimbursed by FEMA, obstacles exist to some state preparedness actions prior to an event (GAO 1993a).

Shortfalls in the Federal response were exacerbated by shortfalls in the state and local responses. As one analyst noted:

> Weaknesses in Florida's state and local emergency training became clear. . . . Most Florida officials acknowledge that more than FEMA was to blame for the slow response to Andrew. State and local officials were hampered by a poor disaster plan, lack of training and poor state–local relations. (Quoted in STFDR 1995)

In a *Miami Herald* opinion piece, Florida Senator Bob Graham (D) criticized the crass political aspects of federal disaster relief (Graham 1992). He argued that Congress and the President consistently "underfund" FEMA to hold down spending in the short run, but also so that politicians "who fight for more money are seen as heroes". Graham argued that if we have a sense of our vulnerability to hurricanes then FEMA could be treated more like an "insurance fund" rather than an "unfunded sham that invites political gimmickry" (cf. GAO 1993b).

In short, the federal government did much to aid victims of Andrew, but at the same time Andrew revealed extensive opportunities to improve the process of federal disaster assistance. The policy challenge that results is to apply the lessons learned in Andrew's aftermath (or any disaster) before the next hurricane strikes.

A test of local government and citizenry

An editorial in *The Miami Herald* noted that the hurricane served as an extreme test of government performance.

> Hurricane Andrew provides a sobering reminder that whom we pick does matter. Before we punch a ballot we should ask ourselves of the candidates: Can I turn to them in the face of disaster and expect them to be there?
> There is a second opportunity here amid the devastation. We in Dade County also have had the rare chance to see the wheels of government turn – or fail to turn – in response to this most severe of tests.

> During the critical first 72 hours, efforts to coordinate the relief programs seemed unavailing. Nobody seemed to be in charge; nobody seemed capable of making decisions that would break up logjams and cause aid to flow. (Fielder 1992)

Breakdowns in the long-term building code process and difficulty in coordinating the relief efforts are examples of interaction between long- and short-term processes that require leadership to be successful. A lesson of Hurricane Andrew is that the people that we elect to public office do affect our lives in tangible ways.

Andrew revealed what is best about society, and, sadly, also what is worst. Many people gave generously of their time and energy to help others prepare for and recover from Andrew. As the hurricane closed in on south Florida, *The Miami Herald* gave thanks to all those who were working furiously to increase public safety during the storm, both during Andrew and over the longer term.

> Thoughts of thanks go not only to planners and building-code enforcers who prepared the way for whatever safety has sheltered South Floridians in this emergency. The debt is owed as well to police, fire, and transportation departments, rumor-control units, medical workers and emergency rescue teams, and the vast army of preparedness, mobilized in a heartbeat even as their own hearts were beating faster for homes and families and loved ones.

Yet looters, opportunists, and criminals showed that not all people are concerned about the welfare of others. Those who sold ice, generators, and food at outrageous prices to those in need remind us again that some people seek to exploit the misfortunes of others for a small, short-term monetary gain. Successful preparedness actions can help limit the opportunities for ill-gotten gains by providing needed materials in the aftermath of a disaster. In addition, preparedness efforts can also limit opportunities for looting and criminal activity following a storm.

7.4.2 Conclusion: restoration

The clean-up from Andrew took months, and was not completed in some areas as late as 1995. Andrew left in its wake an enormous amount of rubble and trash: "Enough debris was hauled away to erect a pile three feet tall and three feet wide – and extend it for 188 miles or from Miami clear across the Everglades to Naples" (Anderson, Lim and Merzer 1992). Yet, for the most part, by the end of 1995 many areas in Dade County were close to fully restored. The following excerpt from a *Miami Herald* analysis of the health of the south Florida economy was published in January 1994, 17 months after Andrew.

Relative to the rest of the country, much of which is still waiting for significant improvement, Florida's economy shined last year [1993]. It led the country in the net number of new jobs created, adding workers at the pace of 16,000 a month. . . . It's a sharp contrast to the perpetual cycle of bad news that Florida, particularly South Florida, faced just a couple of years ago. Layoffs, dropping property values and a sharp drop in new arrivals fed on one another. Unemployment soared into the double digits in Dade and Palm Beach Counties. (Fields 1994, cf. Strouse 1992)

The analysis argued that Andrew helped to turn around south Florida's stagnant economy. "Slowed population growth caused a veritable depression in construction. But Hurricane Andrew generated opportunities" (Fields 1994). In January 1994, according to Florida's Lieutenant Governor Buddy MacKay, the South Florida "community is stronger than it was in 1991" (Fields 1994).

CHAPTER 8

Tropical Cyclone Fundamentals

8.1 FROM KNOWLEDGE TO ACTION

One of the most important consequences of extreme hurricanes like Andrew will likely be the lessons learned regarding the strengths and weaknesses of existing societal responses to hurricanes. In general, however, the identification of lessons and their application are difficult challenges. That difficulty was exposed following a workshop held in the aftermath of Andrew to develop recommendations for Dade County building codes and building code enforcement (FDCA 1992). Securing recommended changes has proven difficult.

> You might think local governments would see this risk and hurry to mend their ways to better protect their citizens. But you would be wrong. In April 1993 – more than seven months after Hurricane Andrew – the Dade County (Florida) Metro Commission voted to postpone making the recommended improvements in its code and enforcement techniques. And almost 10 years after Hurricane Alicia, we are still trying to persuade Texas legislators to adopt the latest building code, never mind enforce it. (Mulady 1994)

The lesson drawn here is that following Andrew, even with numerous clear and specific lessons, drawn and publicized in numerous reports and assessments, effective policy change is difficult to achieve. The challenge is to use the lessons from Andrew and other disasters to improve preparedness in Dade County and elsewhere, before similar tragedies occur. Future Andrews will certainly strike, the central policy question that must be addressed is: How can we reduce our vulnerability through becoming better prepared?

One way to improve preparedness would be for agents of change to put forth a well thought out plan in the immediate aftermath of a disaster. As we have seen, a disaster or extreme event opens a "window of opportunity" for change (Ungar 1995). The window does not stay open for very long – soon follows "a prolonged limbo – a twilight realm of lesser attention or spasmodic recurrences of interest" (Downs 1972). Consequently, those with an interest in improving policy outcomes with respect to hurricanes ought to have a plan of action ready for when conditions do become favorable for policy change. This means that efforts of hurricane policy advocates will be enhanced with an ability to recognize and capitalize on a window of opportunity. Accurate and useful knowledge of hurricanes and society is a necessary element of a rational plan put forward at the right time, to those with authority and interest to act. In the context of the hurricane problem facing the United States, such knowledge includes the collective wisdom from the past which allows for the development and application of alternative actions in the future. Yet the following pattern seems to repeat all too often in the aftermath of every hurricane's impact: general lessons for coping with hurricanes are drawn, but are soon forgotten, only to have to be relearned by another community (and sometimes the same community) in the aftermath of the next hurricane. The difficulty in learning lessons was vividly underscored in the aftermath of Hurricane Hugo (1989) when land developers and home owners rebuilt in vulnerable locations (Seabrook 1990). Leadership is also an important factor in the process of moving from knowledge to action.

With these ideas in mind we have distilled a list of "ten important lessons of hurricanes." We do not expect that every hurricane expert will agree with our list; indeed, it would be healthy for the policy process for such a list to be debated and refined. What the list provides is a summary of important aspects of the hurricane problem that are either generally unknown or underappreciated outside the small community of scholars and decision-makers who focus their efforts on reducing society's vulnerability to hurricanes. Based on society's extensive experience with hurricanes, these fundamentals ought to be widely understood and should be reflected in any plan put forth to reduce society's vulnerability to hurricanes. However, it seems that at times they are variously forgotten, ignored, or belittled. Our purpose in raising these 10 fundamentals is to encourage their adoption in any plan put forward to improve hurricane preparedness.

8.2 TEN IMPORTANT LESSONS OF HURRICANES

1. Tropical cyclones are the most costly natural disaster in the United States (and worldwide)

Available evidence leads to the conclusion that, on a worldwide basis, tropical cyclones are the most costly natural disaster in terms of both economics and casualties. As documented in Chapter 5 and repeated here, in the United States alone, after adjusting for inflation to 1995 dollars, tropical cyclones were responsible for an annual average of $1.6 billion for the period 1950–1989, $2.2 billion over 1950–1995, and $6.2 billion over 1989–1995. China suffered an average of $1.3 billion (unadjusted) in damages related to typhoons over the period 1986–1994. Significant tropical cyclone damages are also experienced by other countries including those in southeast Asia (including Japan, China, and Korea), along the Indian ocean (including Australia, Madagascar, and the southeast African coast), islands of the Caribbean and Pacific, and in Central America (including Mexico). While a full accounting of these damages has yet to be documented and made accessible, it is surely in the billions of dollars, with a reasonable estimate of about $10 billion annually (1995 $). Other estimates range to $15 billion annually (Southern 1992). Experts have estimated that, worldwide, tropical cyclones result in approximately 12 000 to 23 000 deaths. Tropical cyclones have been responsible for a number of the largest losses of life due to a natural disaster. For instance, in April 1991, a cyclone made landfall in Bangladesh resulting in the loss of more than 140 000 lives and disrupting more than 10 million people (and leading to $2 billion in damages). A similar storm resulted in the loss of more than 250 000 lives in November 1970. China, India, Thailand, and the Philippines have also seen loss of life in the thousands in recent years.

2. Hurricane damages in the United States have risen dramatically during an extended period of hurricane quiescence

In recent decades hurricane damages have increased rapidly in the United States. This increase has almost entirely taken place during an extended period of decreasing frequencies of intense hurricanes (Landsea et al. 1996). This means that fewer storms are responsible for the increased damages, and these storms are, on average, no stronger than those of past years. Rather than the number of and strength of storms being the primary factor responsible for the increase in damages, it is the rapid population growth and development in vulnerable coastal locations (Pielke and Landsea 1997). Society has become more vulnerable to hurricane impacts.

The trend of increasing losses during a relatively quiet period of hurricane frequencies should be taken as an important warning. When hurricane frequencies and intensities return to levels observed earlier this century, then losses are sure to increase to record levels unless actions are taken to reduce vulnerability.

3. A large loss of life is still possible in the United States

Inhabitants along the US Atlantic and Gulf Coasts are fortunate in that hurricane watches and warnings are readily available, as are shelters and well-conceived evacuation routes. However, this should not give reason for complacency – the hurricane problem cannot be said to be solved. Disaster planners have developed a number of scenarios that result in a large loss of life here in the United States. For instance, imagine a situation of gridlock as evacuees seek to flee the Florida Keys on the only available road. Or imagine New Orleans, with much of the city below sea level, suffering the brunt of a powerful storm, resulting in tremendous flooding to that low-lying city. Scenarios such as these require constant attention to saving lives. Because the nature of the hurricane problem is constantly changing as society changes, the hurricane problem can never be said to be solved.

4. Tropical cyclone forecasts (seasonal, intensity, and track) can continue to improve; however, societal benefits associated with them depend upon using them effectively

Improved forecasts of tropical cyclone tracks and intensities may, with other actions, contribute to reduced vulnerability by decreasing "overwarning" and allowing for improved decision-making. As scientists note, "to achieve [the benefits of improved forecasts], a program for public awareness and preparedness must be combined with earlier warnings" (HRD 1994). Achieving benefits from improved forecasts is often a challenging task: "as we improve existing forecasts, especially of extreme meteorological events, it is not enough to produce and disseminate a forecast of a specific weather event, but there is a need to improve society's understanding and use of that forecast" (USWRP 1994). Because numerous factors contribute to any particular decision related to societal vulnerability to hurricanes "assessing the economic value of forecasts is not a straightforward task" (Murphy 1994). A forecast is one of several factors which influence a particular (potential) user (cf. Torgerson 1985); factors external to a forecast could enhance or constrain its use. As forecasts improve, attention needs to be paid to improving their use by decision-makers.

Hence, it may be worthwhile for producers of forecasts to ensure that an ongoing parallel research effort is undertaken, targeted at actual and potential users. For instance, if a locale does not have an effective evacuation plan, then an improved forecast would likely make little difference. A parallel program could focus on various decision processes (Chapter 2) to identify opportunities for and constraints on proper and improved use of hurricane forecasts. Counties, states, or SLOSH basins would be appropriate levels of analysis for such a program.

Similarly, as scientists seek to develop and refine reliable seasonal and longer-term forecasts, attention must also be paid to the use/value of these capabilities. The forecasts issued by William Gray are widely disseminated and receive significant attention, yet little, if anything, is known about their usefulness to decision-makers. To determine the usefulness or value to society of such forecasts a series of assessments might be structured using the methodology applied by Glantz (1977, 1979). One such study sought to determine what agricultural decision-makers in Canada would do with a perfect climate forecast one year in advance. Although such a forecast will never be available, an assessment of the impact of a perfect forecast can be useful in determining the value of a less than perfect, but feasible, long-range forecast for the Prairie Provinces. It will also enable us to examine options that various types of decision-makers, from Provincial and Federal government officials to individual farmers, might have to minimize the impact of weather anomalies on agricultural output (Glantz 1977). Glantz (1982) considered the problem in another way and assessed the social costs of an inaccurate forecast. Glantz (1986) concluded from these studies that "the formulation, promulgation, and implementation of a forecast must be carefully assessed, almost on a case by case basis, in order to determine its true value to society". Such assessments of use, misuse, and non-use of forecasts could illuminate the sensitivity of various decision-making processes to improved meteorological products, and to define the upper and lower limits on the value of improved long-term forecasts.

To our knowledge, no such assessment of the value of a long-term hurricane forecast has been conducted. Assessments of the potential and actual use and value of long-term forecasts of hurricane activity would be a valuable contribution to understanding the opportunities for and limitations on actions focused on reducing societal vulnerability to hurricanes. Furthermore, in an era where science is increasingly called upon to demonstrate societal benefits, such assessments have potential to explore how to best leverage investments in research for practical ends.

5. The climate varies on all measurable time scales

To better understand the concepts of climate and weather, imagine yourself sitting at a blackjack table. The range of all possible hands that you might be dealt is analogous to climate; while you do not know what you will get, you do know what to expect. Weather, by contrast, is the hand you are dealt; it is what you get. With a standard deck of 52 cards and repeated experience at the game you might actually be able to play well. With statistical analysis (like card counting) you could perhaps play very well. If climate were constant, analogous to a 52-card deck, then decision-makers in society could also get pretty good. In fact, many decision-makers start off with an assumption that the frequency, intensity, and distribution of climate phenomena are in fact constant. Unfortunately for the needs of policymaking, that simply is not the case.

Imagine that someone is changing the composition of the deck and you do not know exactly how it is changing. They could be adding or removing a few cards, or even replacing cards. As you are dealt individual hands it will be difficult to detect what changes have been made, or in many cases even that changes *have* been made. What has changed is the frequency with which certain hands will appear as well as the relative weight of certain hands. Changes in climate are analogous. For instance, the climate may enter a state in which the formation of hurricanes is more probable than it was in a previous state. Or for particular locations, precipitation might increase or decrease, or thunderstorms might form more often. Such variations in weather statistics have been observed all over the world. Climate does change.

Climate researchers seek to understand why climate changes. In terms of the blackjack analogy, they are trying to understand changes to the deck in order to better anticipate certain outcomes. Because the climate system is large and complex, it is difficult to discern the sources of change. Nonetheless, the scientific community has made notable progress on a number of fronts. For instance, scientists have discovered that increases in sea surface temperature in the eastern and central equatorial Pacific, called El Niño (EN), and differences in atmospheric pressure between the western and eastern Pacific (usually measured between Darwin, Australia and Tahiti), called the Southern Oscillation (SO), are related to various climate phenomena around the world, especially in the tropics. Hurricane frequency in the Atlantic is one phenomenon associated with ENSO events. An understanding of climate variability is essential to effective preparation for future climate impacts, the timing and impacts of which will, to some degree, always be somewhat uncertain. Improved understanding of climate can help to reduce that uncertainty.

6. Recent trends in hurricane frequencies and intensities are not evidence of global warming, and there is considerable reason to prepare better for hurricanes independent of concern about global warming

The phrase "global warming" refers to the possibility that the Earth's climate may change because human activities are altering the composition of the atmosphere. Scientists first raised this possibility more than a century ago, and in recent decades policymakers have begun to express concern about the possibility of climate change. Possible changes that have been discussed in the context of global warming include increasing or decreasing tropical cyclone activity, increased spread of infectious diseases, change in mean global temperature and regional and local temperature variability, and a more active hydrological cycle, including the possibility of more floods (see IPCC 1996a and 1996b for discussion of the science and impacts associated with climate change).

One result of scientific and political concern about the possibility of global warming is a frequent association by the media and the public between it and every extreme weather event. For instance, the cover of *Newsweek* from 22 January 1996 carried the following title: "THE HOT ZONE: Blizzards, Floods, and Hurricanes: Blame Global Warming". This article, and others like it, carries the implication that global warming is responsible for recent climatic extremes. The reality is that it is essentially impossible to attribute any particular weather event to global warming. At the regional level scientists have documented various increasing and decreasing trends in the frequency or magnitude of extreme events, but are not able to associate those changes to global warming. Globally it is difficult for scientists to discern any trends in recent patterns of extreme events (see Chapter 2). Does the lack of a linkage between global warming and extreme weather events mean that the public and policymakers need not concern themselves with climate change? On the contrary, there are many reasons for the public and policymakers to have an increasing concern about the impacts of extreme events, and this concern is independent of the global warming hypothesis.

Given the extensive social and demographic changes since the period of relatively high hurricane incidence earlier this century, it is not only important but imperative to ask whether our current hurricane preparation and response strategies are adequate should the high incidence of landfall along the Atlantic and Gulf coasts return. Simply based on the fact that during various periods in the historical record, hurricane numbers, intensities, and landfall frequencies have been both depressed as well as elevated from the long-term average in the historical record, one can assume that they will eventually rise again. It is therefore prudent to consider current hurricane-related policies under the conditions of past climates and present demographics. This approach is generally called the "forecasting by analogy" method of analysis. The results are frightening.

Research suggests tens of billions of dollars in costs if a major hurricane were to strike a major US metropolitan area: $26 billion for New Orleans, Louisiana; $43 billion for Galveston–Houston, Texas (Chapter 2). Estimates of $50 billion or more in damages resulting from a single storm are becoming commonplace. The fact is that either Hugo and Andrew, as bad as they were, could have been very much worse had they not made landfall over areas that were lowly populated compared to nearby locales. There is reason to believe that Andrew was only 10 miles away from being a $50 billion storm (Chapter 7).

Before asking if we are prepared for the future, we ought to ask if we are prepared even for past known events and climate fluctuations. The future is uncertain – the recent past, however, is certain. Once we consider ourselves "prepared for the past", so to speak, we can seek additional proactive improvements for the future. It would be tragic to ignore the qualitative as well as quantitative base of experience that is readily available. Besides, for many public and private decision-makers the cause of increased hurricane incidence is of less importance than whether the incidence of such extreme meteorological events will increase. While analysis based on global warming remains inconclusive, history tells us with some degree of certainty that the incidence of hurricanes will eventually increase. What will we do if the next several decades were to witness the hurricane activity of the 1940s and 1950s?

The historical record shows that, in the twentieth century, an average of two major hurricanes have struck the US coast every three years. Based on the historical record, if the average damage due to each major hurricane were $4.5 billion, then the US would suffer $3.0 billion in hurricane-related losses per year. Hurricanes Hugo and Andrew suggest that an estimate of $4.5 billion per major storm may be too low. If the average damage due to a major hurricane is instead $7.5 billion, for example, then the annual damages suffered by the US would be at least $5.0 billion even neglecting minor hurricanes and the case of a higher-than-average incidence of landfalling hurricanes. Are we even prepared for the historical average, much less the worst case scenario?

Is there risk in "preparing for the past"? The worst case scenario may be that we overprepare for hurricanes. But, due to the increased exposure of US coastal locations due to demographic shifts, many actions to prepare better for hurricanes could be taken before we become overprepared. The alternative is that we ignore the past and focus on knowing the future, and in the process miss the most important and reliable information available to improve our hurricane preparedness and response strategies.

7. Tropical cyclone landfalls highlight the existing level of societal preparedness

In 1992, Hurricane Andrew served as a dramatic assessment of Dade County's level of exposure to hurricane winds. Figure 7.3, based on an analysis conducted by *The Miami Herald*, shows winds speeds of Hurricane Andrew as well as levels of damage due to the storm in Dade County neighborhoods. The figure is not very precise in its depiction of either wind speed or damage (see Wakimoto and Black 1994 for greater detail on destructive winds and *The Miami Herald* 20 December 1992 for greater detail on damages). Nevertheless, the central message that it depicts is valid: while extreme damage occurred, as might be expected, in the areas of greatest wind speeds, extreme damage also occurred outside the area of greatest wind speeds, which was not expected.

With building codes widely regarded as among the toughest in the nation, why did Dade County see such extreme damage in unexpected areas? A number of studies conducted in the aftermath of Andrew found that, in spite of the tough codes on the books, implementation and enforcement of the codes were often not adequate. Hence, Dade County was more exposed to hurricanes than it might have been (Chapter 7). In one sense, Andrew served as a ruthless assessor of the level of exposure of Dade County to an intense hurricane. In its aftermath, vulnerable areas were visibly and viscerally apparent.

If we are to identify those actions needed to improve a community's preparation for a hurricane's impacts, then we must focus attention on ways to ascertain a community's exposure before a hurricane strikes. This means that we must support efforts to grapple with the messy and challenging task of assessment of hurricane preparedness.

A recent example of an effort to improve assessments of preparedness is provided by the insurance industry. Following Hurricane Andrew the insurance industry learned that a successful building code to reduce storm damages was as much a matter of effective implementation (compliance and enforcement) as it was having a strong code on the books. Thus, one insurance group has begun to evaluate building codes according to the level of implementation (e.g. through enforcement budget, frequency and quality of inspections) rather than simply through the words of the code (BTFFDR 1995). In this manner a more accurate assessment is possible of the health of this particular preparedness process.

8. Short-term decisions are based upon decision processes developed over the long term

When a hurricane approaches a particular stretch of coastline, effective decision-making depends upon the existence of plans, procedure, and prior preparation for the event. In the few hours before a storm strikes, only so much can be done; thus time and efficiency are critical. Lacking effective long-term preparation, a community may well be more vulnerable to a storm's impact. Consider that up to $6 billion of Andrew's total damages were attributed to failure to comply with existing building codes (Noonan 1993). In addition, the successful evacuation of 750 000 residents out of the path of Andrew is largely attributable to the existence of an updated evacuation plan. As obvious as these points seem with hindsight, it is often the case that prior planning for extreme events is overlooked or forgotten as it is not part of "normal" decision routines. Most locations along the US Atlantic and Gulf coasts have not experienced a direct strike from a hurricane in recent memory. Appendix E reprints a guide, put together by officials in Lee County, Florida, to 176 short-term decisions that must be made by a county emergency management office. The guide is instructive because each of the 176 decisions is in some way related to a decision process developed over the long term. Such a "map" of decision for any particular decision-maker is a critical element in evaluating preparedness.

9. Better knowledge of hurricanes, by itself, is generally not sufficient for behavior change

Those who study public response to natural disasters are well versed in the persistence of the fallacy that better information is sufficient to lead to improved decisions. As Sims and Baumann (1983) note in the context of public response to natural disaster: "it doesn't necessarily follow that because information is given it is received or because education is provided there is learning". Knowledge does not always lead to action, but only "under highly specified conditions, and if properly executed, with certain target publics, information *may* lead to awareness and awareness *may* lead to behavior" (ibid, emphasis in original). The same phenomenon of knowledge not leading to action seems to occur not only with regard to public response to natural disasters but with policymakers as well.

Social scientists have explored many of the reasons why "policy happens" and have developed a robust literature (e.g. Olson and Nilson 1982). From this body of theory and practice one point stands clear: knowledge of a problem (or a risk, threat, or danger) does not inevitably lead to effective policy action. History, however, shows that a disaster is one factor which can lead to policy action.

According to Hilgartner and Bosk (1988) drama, novelty and saturation, and culture and politics also influence what becomes defined as an important social problem and what does not. In the absence of an extreme event that mobilizes political action, policy for reducing a community's vulnerability to hurricanes must meet several criteria in a business-as-usual environment (Nilson 1985). First, the threat must be demonstrated. Second, potential responses must be shown to have a significant likelihood of being effective. Third, policy options must not be viewed as imposing excessive costs or changes on the community.

10. Society knows, in large part, how to respond to hurricanes

One of the most frustrating aspects of society's response to hurricanes (and natural hazards more generally) is the realization that in many cases society currently knows enough to take effective actions to reduce its vulnerabilities. For instance, more than 17 years before Andrew made landfall south of Miami, White and Haas (1975) presciently described the event as a "future disaster", that is "human suffering and economic disruption which will result from events whose coming is certain but whose timing is completely uncertain" (p. 29). While no one could have predicted exactly when a hurricane would strike a particular section of the coast, for the most part these researchers accurately anticipated the event, its impacts, and the shortfalls in societal response that would be identified in the storm's aftermath.

Just as White and Haas identified Miami, Florida as the site of a future hurricane disaster, it is certain that countless other coastal communities, large and small, will suffer the impact of hurricanes in coming years. The next major hurricane like Andrew might be an Alex, Charley, or Ivan in 1998, or Debby, Isaac, or Keith in 2000. It may not occur for years, but it will happen. As the number of people and property at risk to hurricanes continues to rise, the potential for extreme impacts grows. The *Natural Disaster Survey Report* conducted in Andrew's aftermath concluded: "That another 'Andrew' will occur is not conjecture, it is a certainty. Our ability to respond to that reality hinges on how we answer the call to mitigate" (DOC 1993).

North America has over 500 years of *recorded* history of hurricane impacts. Society has over the centuries developed a large body of experience about how to reduce its vulnerability to hurricanes. We have come a long way since up to 12 000 people died in the Galveston storm of 1900. Today, we know what hurricane preparedness entails. A primary challenge is the application of that knowledge in specific present day settings as well as refinement of existing response strategies (e.g. through the use of improved hurricane forecasts).

Although the processes of preparedness for hurricanes are generally well understood, how to actually translate that knowledge into lowered vulnerability is still neither well understood nor well implemented.

8.3 LAST WORDS

In conclusion, we offer three recommendations to help move hurricane policy from knowledge to action. While we focus these recommendations on the United States, we believe that they have broad relevance for other countries that suffer the impacts of tropical cyclones.

1. Form a hurricane policy team to provide leadership

In the United States, hurricanes present a policy problem for more than just the directly impacted communities of a particular storm. Hurricanes have local, regional, and at times national impacts (e.g. on insurance). In addition, the lessons learned in one community following a storm's impact might be relevant to improving hurricane preparedness of another community in a different state or region. With vulnerability to hurricanes rising in the United States, impacts will likely increase and the lessons of experience will likely become commensurately more important.

For these reasons, we recommend the formation of a hurricane policy team to provide leadership and guidance to federal, state and local, public and private decision-makers. The team's composition should be geographically diverse. In addition, to be most effective, this team should include people with expertise in as many of the facets of the hurricane problem as is practical. For instance, engineering, emergency management, evacuation, land use, insurance, forecasting, elected officials, and the general public should all be represented on the team. The team could be initiated by the Federal government, in a fashion similar to earthquake and flood policy teams now in existence, but could also be initiated by a regional association of governors, emergency managers, or others. The mission of the team would include the provision of evaluations on the relative national, regional, and local vulnerability and to provide guidance and leadership in the aftermath of a storm's impact, both to the directly affected community and those far removed.

2. Assess community vulnerability

One task that a hurricane policy team could embark upon immediately would be to assess in a systematic and authoritative manner the nation's vulnerability to hurricane impacts. We have argued the need for such assessments throughout this book and have offered some suggestions as to important factors that might be included in such assessments.

A vulnerability assessment would provide local, state, and national decision-makers with important information on both a community's relative state of preparedness, as compared with other communities, as well as a sense of the nation's total vulnerability to hurricanes. Such an assessment is important from the standpoint of providing relevant information to the process of reducing a particular community's vulnerability. Typically, a community's vulnerability is revealed only by a hurricane's impact, which for them is too late. Vulnerability assessment is also essential to the process of effective allocation of scarce resources in this area. It would help decision-makers to allocate resources between enforcement of building codes versus, for instance, evacuation planning. Also decision-makers currently have little systematic knowledge as to whether Wilmington, North Carolina is more vulnerable to hurricane impacts than Atlantic City, New Jersey. Additionally, little systematic information is available as to the relative importance of vulnerability to hurricanes versus vulnerability to other disasters such as floods or earthquakes.

3. Keep ready an updated plan of action for when a window of opportunity for change arises

A hurricane policy team could also provide leadership and sensible guidance when a window of opportunity for policy action arises in the aftermath of a disaster. The team could bring to local, regional, and national decision fora well-thought-out and politically tested policy alternatives to reduce vulnerability to hurricanes. These recommendations would of course have to be tailored to the particular context of each decision as there will not be one solution that works equally well everywhere. However, a leadership team can work to ensure that lessons of the past are not forgotten in the pressures of the moment and that an authoritative, independent voice is available to help a community (or more broadly the nation) to avoid recreating those conditions which led to the disaster. The hurricane policy team could also be responsible for evaluating a community's response to and recovery from a hurricane, in the process providing additional knowledge and lessons to the body of experience with hurricanes.

Tropical cyclones on planet Earth are a constant reminder of the power of nature. In a few hours, a powerful storm can devastate one of our most advanced cities. As a consequence of population growth and related development in exposed coastal locations, we have arguably become more vulnerable to hurricanes than at any time in the recent past. For some time, scientific and technological developments such as the weather satellite have provided an illusion that the hurricane problem was slowly, but systematically, becoming less and less important. Today we have a richer understanding of the hurricane problem. Science and technology have indeed provided society with great benefits with respect to the hurricane problem, but by no means is the problem going away. Society continues to increase its exposure at a rapid rate. In the Atlantic Basin, measures of increased property losses are one indication of increased exposure, but miss the fact that hurricane incidence has been depressed in recent decades. It is likely that the trend of constantly decreasing deaths each decade associated with hurricanes in the US will not continue. Hurricane impacts, in the Atlantic basin and beyond, are certain to become a more important policy issue in the near future than they have been in the recent past. As long as societal vulnerability to hurricanes continues to increase, we will continue to see damages and casualties increase, making reduction of vulnerability all the more important.

APPENDIX A

Additional Reading

CHAPTER 1

Byerly, R., 1995: U.S. science in a changing context: A perspective. In *U.S. National Report to the International Union of Geodesy and Geophysics 1991–1994*, R.A. Pielke, Ed., American Geophysical Union, Washington, DC, A1–A16.

Douglas, M.S., 1947: *The Everglades: River of Grass*, Banyan Books, Miami, FL, 447 pp.

Dunn, G.E. and B.I. Miller, 1964: *Atlantic Hurricanes*, Louisiana State University Press, Baton Rouge, LA, 377 pp.

Fernandez-Partagas, J., 1995: The deadliest Atlantic tropical cyclones, 1492–1994. NOAA Technical Memorandum, NWS-47, 41 pp.

Fisher, D.E., 1994: *The Scariest Place on Earth: Eye to Eye with Hurricanes*, Random House, New York, 250 pp.

Hurston, Z.N., 1969: *Their Eyes were Watching God*, Negro Universities Press, New York, 286 pp.

Lodge, T.E., 1994: *The Everglades Handbook: Understanding the Ecosystem*, St Lucie Press, 228 pp.

Ludlam, D.M., 1963: *Early American Hurricanes: 1492–1870*, American Meteorological Society, Boston, MA, 198 pp.

Machalek, J., 1992: *A Bibliography of Weather and Climate Hazards*, 2nd Edition. Topical bibliography No. 16. Natural Hazards Research and Applications Information Center, University of Colorado, Boulder, CO, 331 pp.

Pielke, Jr., R.A. and M.H. Glantz, 1995: Serving science and society: Lessons from large-scale atmospheric science programs. *Bull. Am. Meteor. Soc.*, 76, 2445–2458.

Rappaport, E.N., P.J. Hebert, J.D. Jarrell, and M. Mayfield, 1996: The deadliest, costliest, and most intense United States hurricanes of this century (and other frequently requested hurricane facts). NOAA Technical Memorandum NWS TPC-1, February 1996, National Hurricane Center, Miami, FL.

Sarewitz, D., 1996: *Frontiers of Illusion: Science, Technology and the Politics of Progress*, Temple University Press, Philadelphia, PA.

Simpson, R.H. and H. Riehl, 1981: *The Hurricane and its Impact*, Louisiana State University Press, Baton Rouge, LA, 398 pp.

Tannehill, I.R., 1952: *Hurricanes: Their Nature and History*, 8th edition, Princeton University Press, Princeton, NJ.

CHAPTER 2

Alexander, D., 1993: *Natural Disasters*. Chapman and Hall, New York.

Bardwell, L.V., 1991: Problem-framing: A perspective on environmental problem solving. *Envir. Manage.*, 15, 603–612.

Burton, I., R.W. Kates, and G.F. White, 1993: *The Environment as Hazard*, 2nd edition, Guilford Press, New York, 290 pp.
Cuny, F.C., 1983: *Disasters and Development*, Oxford University Press, New York, 278 pp.
Dewey, J., 1933: *John Dewey: The Later Works, 1925–1953*. Carbondale and Edwardsville, Southern Illinois University Press, 113–139.
Forester, J., 1984: Bounded rationality and the politics of muddling through. *Public Admin. Rev.*, **44**, 23–31.
Hewitt, K., 1983: *Interpretations of Calamity from the Viewpoint of Human Ecology*, Allen and Unwin, Boston, MA, 304 pp.
Wildavsky, A., 1979: *Speaking Truth to Power: The Art and Craft of Policy Analysis*, Little, Brown and Co, Boston, MA.
Winchester, P, *Power, Choice, and Vulnerability: A Case Study in Disaster Mismanagement in South India, 1977–1988*, James and James, London, 225 pp.

CHAPTER 3

Anthes, R.A., 1982: *Tropical Cyclones: Their Evolution, Structure and Effects*, American Meteorological Society, Boston, MA, 208 pp.
Dunn, G.E. and B.I. Miller, 1964: *Atlantic Hurricanes*, Louisiana State University Press, Baton Rouge, LA, 377 pp.
Elsberry, R.L., W.M. Frank, G.J. Holland, J.D. Jarrell, and R.L. Southern, 1987: A global view of tropical cyclones. Publication financed by a grant from the Office of Naval Research, Marine Meteorology Program, Robert F. Abbey, Director, 185 pp.
Emanuel, K.E., 1996: Maximum hurricane intensity estimation. On the World Wide Web: http://cirrus.mit.edu/~emanuel/pcmin/pclat/pclat.html
Gray, W.M., 1981: Recent advances in tropical cyclone research from rawinsonde composite analysis. World Meteorological Organization, 407 pp.
Gray, W.M., 1997: Tropical cyclones. World Meteorological Organization, 194 pp. (in draft).
Kotsch, W.J., 1977: *Weather for the Mariner*, Naval Institute Press, Annapolis, MD, 272 pp.
Landsea, C., 1996: Frequently asked questions. Found at http://tropical.atmos.colostate.edu/text/tcfaq1.html
Merrill, R.T., 1984: A comparison of large and small tropical cyclones. *Monthly Weather Review*, **112**, 1408–1414.
Montgomery, M.T. and B.F. Furrell, 1993: Tropical cyclone formation. *J. Atmospheric Science*, **50**, 285–310.
Neumann, C.J., B.R. Jarvinen, C.J. McAdie, and J.D. Elms, 1993: Tropical cyclones of the North Atlantic Ocean, 1871–1992. National Environmental Satellite, Data, and Information Service, National Climatic Data Center, Asheville, N.C. 28801, 193 pp.
NOAA, 1993: *"Hurricane!" A Familiarization Booklet*, US Department of Commerce, NOAA PA 91001, 36 pp.

CHAPTER 4

DeMaria, M., 1996: The effect of vertical shear on tropical cyclone intensity change. *J. Atmos. Sci.*, **53**, 2076–2087.

Kurikara, Y. and R.E. Tuleya, 1974: Structure of a tropical cyclone developed in a three-dimensional simulation model. *J. Atmospheric Science*, **31**, 893–919.

National Hurricane Center Website at http://www.nhc.noaa.gov

Ooxana, K., 1969: Numerical simulation of the life cycle of tropical cyclones. *J. Atmospheric Science*, **26**, 3–40.

World Meteorological Organization, 1993: *Global Guide to Tropical Cyclone Forecasting*. WMO/TD – NO. 560, Report No. TCP-31, G. Holland, Ed.

CHAPTER 5

Barnes, J., 1995: *North Carolina's Hurricane History*. The University of North Carolina Press, 211 pp.

Bush, D.M., C.A. Webb, R.S. Young, B.D. Johnson, and G.M. Bates, 1996: Impact of Hurricane Opal on the Florida/Alabama coast. Quick response report #84. Found at: http://adder.colorado.edu:80/~hazctr/Home.html

Changnon, S.A., D. Changnon, E.R. Fosse, D.C. Hoganson, R.J. Roth, Sr., and T. Totsch, 1996: Impacts and responses of the weather insurance industry to recent weather extremes. Final Report to the University Corporation for Atmospheric Research, CRR-41, May 1996.

Hebert, P.J., J.D. Jarrell, and M. Mayfield, 1996: The deadliest, costliest, and most intense United States hurricanes of this century (and other frequently requested hurricane facts). NOAA Technical Memorandum NWS TPC-1, February 1996, National Hurricane Center, Miami, FL.

Simpson, R.H. and H. Riehl, 1981: *The Hurricane and its Impact*, Louisiana State University Press, Baton Rouge, LA, 398 pp.

Stone, G.W., J.M. Grymes III, C.K. Armbruster, J.P. Xu, and O.K. Huh, 1996: Researchers study impact of Hurricane Opal on Florida coast. *Eos*, **77**, 181 and 184.

CHAPTER 6

Ayscue, J.K., 1996: Hurricane damage to residential structures: Risk and mitigation. National Hazards Research Working Paper #94. Found at: http://adder.colorado.edu/~hazctr/wp/wp94/wp94.html

Baker, E.J., 1991: Hurricane evacuation behavior. *Int. J. Mass Emergencies and Disasters*, **9**, 287–310.

Hearn Morrow, B. and A. Ragsdale, 1996: Early response to Hurricane Marilyn in the U.S. Virgin Islands. Quick Response Report #82. Found at: http://adder.colorado.edu:80/~hazctr/Home.html

GENERAL

Approximately once a year, the publication *Monthly Weather Review* has a summary of the previous year's hurricane activity. The references for these reports since 1980 are listed below.

Lawrence, M.B., 1981: Atlantic hurricane season of 1980. *Mon. Wea. Rev.*, **109**, 1567–1582.

Lawrence, M.B, 1982: Atlantic hurricane season of 1981. *Mon. Wea. Rev.*, **110**, 852–866.

Clark, G.B, 1983: Atlantic hurricane season of 1982. *Mon. Wea. Rev.*, **111**, 1071–1079.

Case. R.A. and H.P. Gerrish, 1984: Atlantic hurricane season of 1983. *Mon. Wea. Rev.*, **112**, 1083–1092.

Lawrence, M.B. and G.B. Clark, 1985: Atlantic hurricane season of 1984. *Mon. Wea. Rev.*, **113**, 1228–1237.

Case, R.A., 1986: Atlantic hurricane season of 1985: *Mon. Wea. Rev.*, **114**, 1390–1405.

Lawrence, M.B., 1988: Atlantic hurricane season of 1986: *Mon. Wea. Rev.*, **116**, 2155–2160.

Case, R.A. and H.P. Gerrish, 1989: Atlantic hurricane season of 1987. *Mon. Wea. Rev.*, **116**, 939–949.

Lawrence, M.B. and J.M. Gross, 1989: Atlantic hurricane season of 1988. *Mon. Wea. Rev.*, **117**, 2248–2259.

Case, B. and M. Mayfield, 1990: Atlantic hurricane season of 1989. *Mon. Wea. Rev.*, **118**, 1165–1177.

Mayfield, M. and M.B. Lawrence, 1991: Atlantic hurricane season of 1990. *Mon. Wea. Rev.*, **119**, 2014–2026.

Pasch, R.J. and L.A. Avila, 1992: Atlantic hurricane season of 1991. *Mon. Wea. Rev.*, **120**, 2671–2687.

Mayfield, M., L.A. Avila, and E.N. Rappaport, 1994: Atlantic hurricane season of 1992. *Mon. Wea. Rev.*, **122**, 517–538.

Pasch, R.J. and E.N. Rappaport, 1995: Atlantic hurricane season of 1993. *Mon. Wea. Rev.*, **123**, 871–886.

Avila, L.A. and E.N. Rappaport, 1996: Atlantic hurricane season of 1994. *Mon. Wea. Rev.*, **124**, 1558–1578.

Economic and Casualty Data for the United States

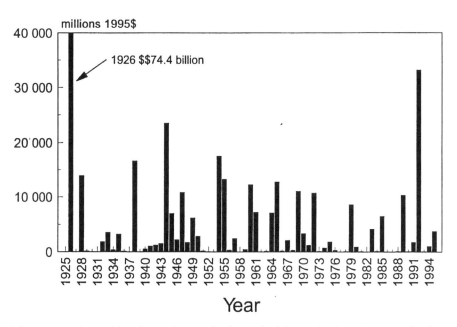

Figure B.1   Annual hurricane damage in the United States 1929–1995, normalized to 1995 values using changes in inflation, wealth, and coastal county populations. The methodology seeks to estimate the damages associated with each hurricane season had it occurred in 1995. See Pielke and Landsea 1997; see also Table B.4.

Table B.1   The 30 deadliest mainland United States hurricanes 1900–1995 (Hebert, Jarrell and Mayfield 1996) and eight other deadly hurricanes

Rank	Hurricane	Year	Category	Deaths
1	TX (Galveston)	1900	4	8000[†]
2	FL (SE/Lake Okeechobee)	1928	4	1836
3	FL (Keys)/S. TX	1919	4	600[‡]
4	New England	1938	3*	600
5	FL (Keys)	1935	5	408
6	Audrey (SW LA/N TX)	1957	4	390
7	NE US	1944	3*	390[∞]
8	LA (Grand Isle)	1909	4	350
9	LA (New Orleans)	1915	4	275
10	TX (Galveston)	1915	4	275
11	Camille (MS/SE LA/VA)	1969	5	256
12	FL (Miami)/MS/AL/Pensacola	1926	4	243
13	Diane (NE US)	1955	1	184
14	SE FL	1906	2	164
15	MS/AL/Pensacola	1906	3	134
16	Agnes (FL/NE US)	1972	1	122
17	Hazel (SC/NC)	1954	4*	95
18	Betsy (SE FL/SE LA)	1965	3	75
19	Carol (NE US)	1954	3*	60
20	SE FL/SE LA/MS	1947	4	51
21	Donna (FL/Eastern US)	1960	4	50
22	GA/SC/NC	1940	2	50
23	Carla (N & Central TX)	1961	4	46
24	TX (Velasco)	1909	3	41
25	TX (Freeport)	1932	4	40
26	S TX	1933	3	40
27	Hilda (Central LA)	1964	3	38
28	SW LA	1918	3	34
29	SW FL	1910	3	30
30	Alberto (NW FL, GA, AL)	1994	TS	30

Pre-1900 or not Atlantic/Gulf Coast

2	LA	1893	Unk	2000
2–3	SC/GA	1893	Unk	1000–2000
3	GA/SC	1881	Unk	700
9	Puerto Rico	1928	4	312
13	USVI, Puerto Rico	1932	2	225
17	Donna (St Thomas, VI)	1960	4	107
24	Southern CA	1939	TS	45
24	Eloise (Puerto Rico)	1975	TS	44

*   Moving at more than 30 miles per hour.
†   May actually have been as high as 10 000 to 12 000.
‡   Over 500 lost on ships at sea; 600–900 estimated deaths.
∞   Some 344 of these lost on ships at sea.
TS,   Only of Tropical Storm intensity.
Unk, Intensity not sufficiently known to establish category.

Table B.2   The 30 costliest mainland United States hurricanes, 1900–1995 (Hebert, Jarrell and Mayfield 1996) and seven other costly hurricanes

Rank	Hurricane	Year	Category	Damage (US $)
1	Andrew (SE FL/SE LA)	1992	4	26 500 000 000
2	Hugo (SC)	1989	4	7 000 000 000
3	Opal (NW FL/AL)	1995	3	3 000 000 000†
4	Frederic (AL/MS)	1979	3	2 300 000 000
5	Agnes (FL/NE US)	1972	1	2 100 000 000
6	Alicia (N TX)	1983	3	2 000 000 000
7	Bob (NC, NE US)	1991	2	1 500 000 000
7	Juan (LA)	1985	1	1 500 000 000
9	Camille (MS/SE LA/VA)	1969	5	1 420 700 000
10	Betsy (SE FL/SE LA)	1965	3	1 420 500 000
11	Elena (MS/AL/NW FL)	1985	3	1 250 000 000
12	Gloria (Eastern US)	1985	3*	900 000 000
13	Diane (NE US)	1955	1	831 700 000
14	Allison (N TX)	1989	TS	500 000 000
14	Alberto (NW FL/GA/AL)	1994	TS	500 000 000
16	Eloise (NW FL)	1975	3	490 000 000
17	Carol (NE US)	1954	3*	461 000 000
18	Celia (S TX)	1970	3	453 000 000
19	Carla (N & Central TX)	1961	4	408 000 000
20	Claudette (N TX)	1979	TS	400 000 000
20	Gordon (S & Central FL/NC)	1994	TS	400 000 000
22	Donna (FL/Eastern US)	1960	4	387 000 000
23	David (FL/Eastern US)	1979	2	320 000 000
24	New England	1938	3*	306 000 000
25	Kate (FL Keys/NW FL)	1985	2	300 000 000
25	Allen (S TX)	1980	3	300 000 000
27	Hazel (SC/NC)	1954	4*	281 000 000
28	Dora (NE FL)	1964	2	250 000 000
29	Beulah (S TX)	1967	3	200 000 000
30	Audrey (SW LA/N TX)	1957	4	150 000 000
Not Atlantic/Gulf Coast				
6	Iniki (Kauai, HI)	1992	Unk.	1 800 000 000
7	Marilyn (USVI, PR)	1995	2	1 500 000 000
12	Hugo (USVI, PR)	1989	4	1 000 000 000
22	Olivia (CA)	1982	TD	325 000 000
23	Iwa (Kauai, HI)	1982	Unk	312 000 000
24	Norman (CA)	1978	TD	300 000 000
29	Kathleen (CA & AZ)	1976	TD	160 000 000

*   Moving at more than 30 miles per hour.
†   Current estimate subject to change.
TS,   Only of Tropical Storm intensity.
TD,   Only a Tropical Depression.
Unk,   Intensity not sufficiently known to establish category.

Table B.3   The 30 costliest mainland United States hurricanes 1900–1995 adjusted for inflation (Hebert, Jarrell and Mayfield 1996) and four other costly hurricanes

Rank	Hurricane	Year	Category	Damage (US $)[†]
1	Andrew (SE FL/SE LA)	1992	4	28 620 000 000
2	Hugo (SC)	1989	4	7 910 000 000
3	Agnes (FL/NE US)	1972	1	6 930 000 000
4	Betsy (SE FL/SE LA)	1965	3	6 875 220 000
5	Camille (MS/SE LA/VA)	1969	5	5 640 179 000
6	Diane (NE US)	1955	1	4 516 131 000
7	Frederic (AL/MS)	1979	3	3 933 000 000
8	New England	1938	3*	3 864 780 000
9	Opal (NW FL/AL)	1995	3	2 880 000 000[‡]
10	Alicia (N TX)	1983	3	2 760 000 000
11	Carol (NE US)	1954	3*	2 549 330 000
12	Carla (N & Central TX)	1961	4	2 072 640 000
13	Donna (FL/Eastern US)	1960	4	1 962 090 000
14	Juan (LA)	1985	1	1 950 000 000
15	Celia (S TX)	1970	3	1 694 220 000
16	Bob (NC/NE US)	1991	2	1 635 000 000
17	Elena (MS/AL/NW FL)	1985	3	1 625 000 000
18	Hazel (SC/NC)	1954	4*	1 553 930 000
19	FL (Miami)/MS/AL/Pensacola	1926	4	1 414 560 000
20	N TX (Galveston)	1915	4	1 264 800 000[∞]
21	Dora (NE FL)	1964	2	1 245 000 000
22	Eloise (NW FL)	1975	3	1 190 700 000
23	Gloria (Eastern US)	1985	3*	1 170 000 000
24	NE US	1944	3*	994 000 000
25	Beulah (S TX)	1967	3	900 000 000
26	N TX (Galveston)	1900	4	759 909 000"
27	SE FL/SE LA/MS	1947	4	756 800 000
28	Audrey (SW LA/N TX)	1957	4	748 500 000
29	Claudette (N/TX)	1979	TS	684 000 000
30	Cleo (SE FL)	1964	2	639 930 000

Not Atlantic/Gulf Coast

15	Iniki (Kauai, HI)	1992	Unk	1 944 000 000
19	Marilyn (USVI, E PR)	1995	2	1 440 000 000
24	Hugo (USVI, PR)	1989	4	1 130 000 000
24	San Felipe (PR)	1928	4	1 071 000 000

*   Moving at more than 30 miles per hour.
†   Adjusted to 1994 dollars on basis of US DOC Implicit Price Deflator for Construction; 1995 damages adjusted downward.
‡   Current estimate subject to change.
∞   Damage estimate was considered too high in 1915 adjustment.
"   Using 1915 cost adjustment base – none available prior to 1915.
TS, Only of Tropical Storm intensity, included because of high damage.
Unk, Intensity not sufficiently known to establish category.

Table B.4   Top 30 Damaging Hurricanes 1925–1995 – Normalized to 1995 dollars by inflation, personal property increases, and coastal county population changes (1900–1995). (Asterisks indicate hurricanes included from years 1900–1924 using simplifying assumptions to extend the normalization methodology to 1900.) Source: Pielke and Landsea (1997). See Figure B.1

Rank	Hurricane	Year	Category	Damage (US$ billions)
1	SE Florida/Alabama	1926	4	72.303
2	ANDREW (SE FL/LA)	1992	4	33.094
3	*N Texas (Galveston)	1900	4	26.619
4	*N Texas (Galveston)	1915	4	22.602
5	SW Florida	1944	3	16.864
6	New England	1938	3	16.629
7	SE Florida/Lake Okeechobee	1928	4	13.795
8	BETSY (SE FL/LA)	1965	3	12.434
9	DONNA (FL/Eastern US)	1960	4	12.048
10	CAMILLE (MS/LA/VA)	1969	5	10.965
11	AGNES (NW FL, NE US)	1972	1	10.705
12	DIANE (NE US)	1955	1	10.232
13	HUGO (SC)	1989	4	9.380
14	CAROL (NE US)	1954	3	9.066
15	SE Florida/Louisiana/Alabama	1947	4	8.308
16	CARLA (N & Central TX)	1961	4	7.069
17	HAZEL (SC/NC)	1954	4	7.039
18	NE US	1944	3	6.536
19	SE Florida	1945	3	6.313
20	FREDERIC (AL/MS)	1979	3	6.293
21	SE Florida	1949	3	5.838
22	*S Texas	1919	4	5.368
23	ALICIA (N TX)	1983	3	4.056
24	CELIA (S TX)	1970	3	3.338
25	DORA (NE FL)	1964	2	3.108
26	OPAL (NW FL/AL)	1995	3	3.000
27	CLEO (SE FL)	1964	2	2.435
28	JUAN (LA)	1985	1	2.399
29	AUDREY (LA/N TX)	1957	4	2.396
30	KING (SE FL)	1950	3	2.266

APPENDIX B 203

Table B.5 Estimated annual deaths and damages (unadjusted and inflation adjusted to constant 1994 dollars) in the mainland United States from landfalling Atlantic or Gulf tropical cyclones 1900–1995 (Hebert, Jarrell and Mayfield 1996)

Year	Deaths	Damage ($ millions) Unadjusted	Adjusted*	Year	Deaths	Damage ($ millions) Unadjusted	Adjusted*
1900	8 000	30	790	1948	3	18	113
1901	10	1	26	1949	4	59	370
1902	0	Minor	Minor	1950	19	36	222
1903	15	1	26	1951	0	2	11
1904	5	2	53	1952	3	3	16
1905	0	Minor	Minor	1953	2	6	33
1906	298	3	79	1954	193	756	4 180
1907	0	0	0	1955	218	985	5 348
1908	0	0	0	1956	19	27	139
1909	406	8	211	1957	400	152	758
1910	30	1	26	1958	2	11	55
1911	17	1	26	1959	24	23	116
1912	1	Minor	Minor	1960	65	396	2 006
1913	5	3	79	1961	46	414	2 101
1914	0	0	0	1962	3	2	10
1915	550	63	1 660	1963	10	12	59
1916	107	33	723	1964	49	515	2 564
1917	5	Minor	Minor	1965	75	1 445	6 996
1918	34	5	71	1966	54	15	70
1919	287	22	278	1967	18	200	900
1920	2	3	30	1968	9	10	43
1921	6	3	38	1969	256	1 421	5 647
1922	0	0	0	1970	11	454	1 699
1923	0	Minor	Minor	1971	8	213	747
1924	2	Minor	Minor	1972	122	2 100	6 924
1925	6	Minor	Minor	1973	5	3	9
1926	269	112	1 415	1974	1	150	396
1927	0	0	0	1975	21	490	1 191
1928	1 836	25	315	1976	9	100	233
1929	3	1	12	1977	0	10	22
1930	0	Minor	Minor	1978	36	20	38
1931	0	0	0	1979	22	3 045	5 210
1932	0	0	0	1980	2	300	463
1933	63	47	701	1981	0	25	36
1934	17	5	68	1982	0	Minor	Minor
1935	414	12	163	1983	22	2 000	2 751
1936	9	2	28	1984	4	66	88
1937	0	Minor	Minor	1985	30	4 000	5 197
1938	600	306	3 864	1986	9	17	21
1939	3	Minor	Minor	1987	0	8	10
1940	51	5	66	1988	6	9	11
1941	10	8	98	1989	56	7 670	8 640
1942	8	27	286	1990	13	57	63
1943	16	17	169	1991	16	1 500	1 637
1944	64	165	1 641	1992	24	26 500	28 687
1945	7	80	773	1993	4	57	59
1946	0	5	41	1994	38	973	973
1947	53	136	935	1995	29	3 723	3 582

* Adjusted to 1994 dollars based on US Department of Commerce implicit Price Deflator for Construction; 1995 damages adjusted downward.

Selected Data on Tropical Storm and Hurricane Incidence in the Atlantic Ocean Basin

This appendix presents various trend data on the climatology of tropical cyclones in the Atlantic Basin.

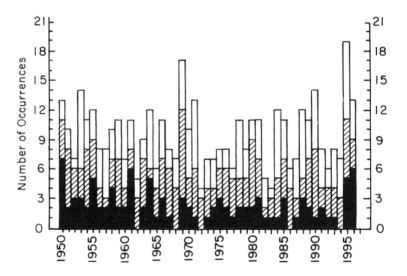

Figure C.1 Number of tropical storms (shaded) and hurricanes (unshaded) in the Atlantic for 1950–1996. Data provided by W. Gray. Graph based on Neumann et al. (1993). Note that data on tropical cyclone occurrences prior to 1950 are considered to be unreliable because some storms escaped detection (C. Landsea, 1997, personal communication)

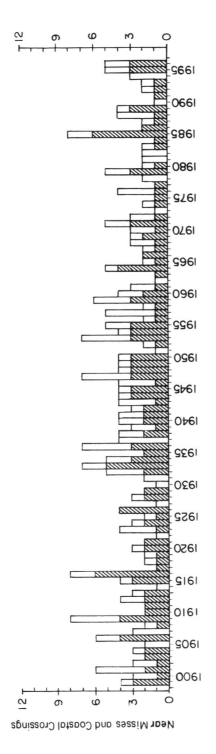

Figure C.2 Number of tropical storms (shaded) and hurricanes (unshaded) that have crossed or passed immediately adjacent to the US mainland 1899–1996. The data are considered reliable over this period. Source: updated from Neumann et al. (1993)

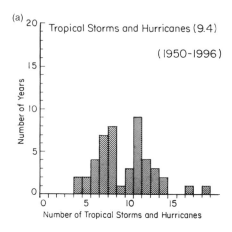

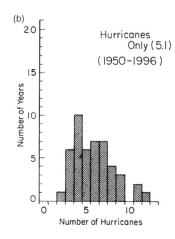

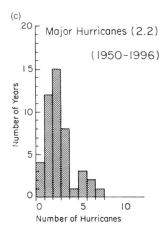

Figure C.3   (a) The frequency of tropical storms and hurricanes in the Atlantic basin for the period 1950–1996. The most active season during this period was 1995 with 19 named storms. The least active seasons were 1972 and 1983 with only four named storms. In each of those years, however, a hurricane made landfall in the US resulting in multi-billion dollar damages: Agnes (1972) and Alicia (1983). While the average annual frequency over this period is 9.4, note that in very few years was the frequency actually 9 or 10. Over this period the basin has been either relatively quiet or somewhat active; there does not appear to be a single "typical" season, but rather two types of "typical" seasons. (b) The frequency of hurricanes only over the period 1950–1996. The most hurricanes was 12 in 1969 and the fewest was 3 in 1962. The annual average is 5.1. (c) The frequency of intense hurricanes over the same period with a maximum of 7 in 1950 and none in a number of years. The annual average is 2.2

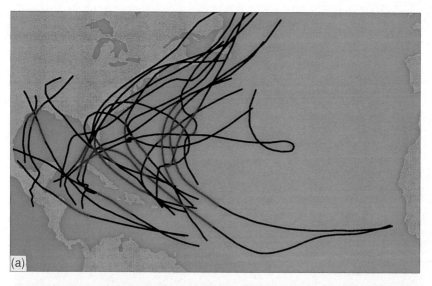

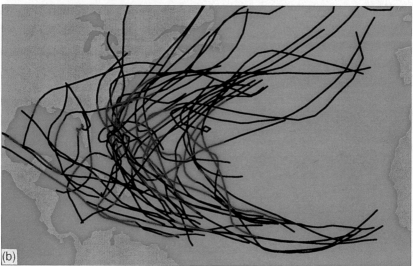

Figure C.4   Tracks of intense hurricanes for (a) 1940 – 1949; (b) 1950 – 1959. The black portion of the storm track indicates that the intensity was less than Category 3 at that point. Red indicates equal or greater than Category 3 strength. Source: National Hurricane Center. Figures prepared by Joe Eastman

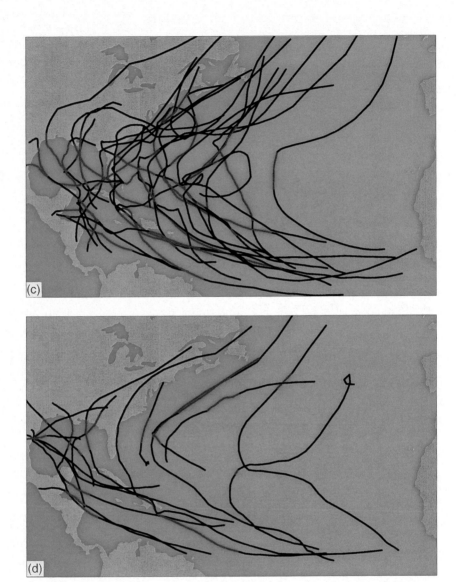

Figure C.4 *Cont:* Tracks of intense hurricanes for (c) 1960 – 1969; (d) 1950 – 1959

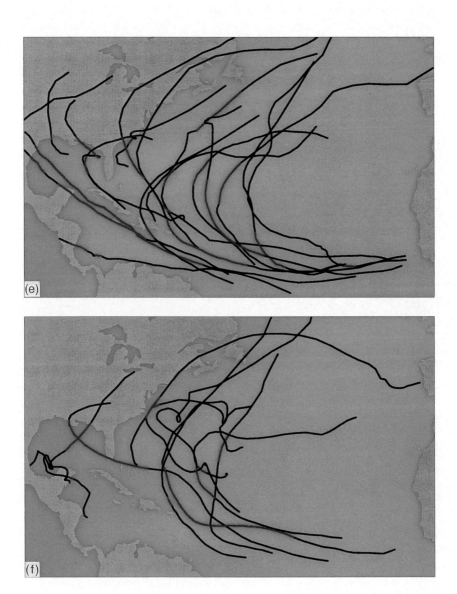

Figure C.4   *Cont:* Tracks of intense hurricanes for (e) 1980 – 1989; (f) 1990 – 1996

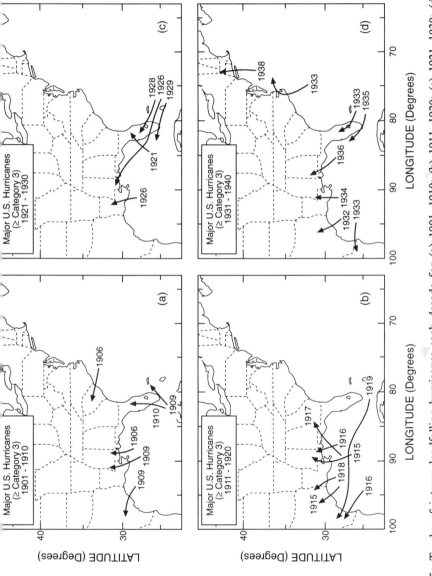

Figure C.5 Tracks of intense landfalling hurricanes each decade for (a) 1901–1910; (b) 1911–1920; (c) 1921–1930; (d) 1931–1940; (e) (*see overleaf*) 1941–1950; (f) 1951–1960; (g) 1961–1970; (h) 1971–1980; (i) 1981–1990; (j) 1991–1996. Source: adapted from Hebert, Jarrell and Mayfield (1996)

208

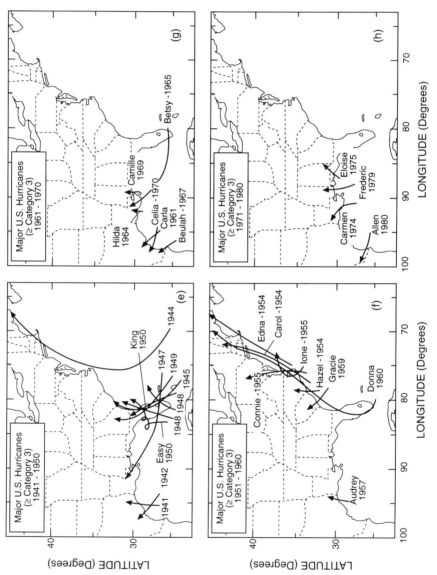

Figure C.5 (continued)

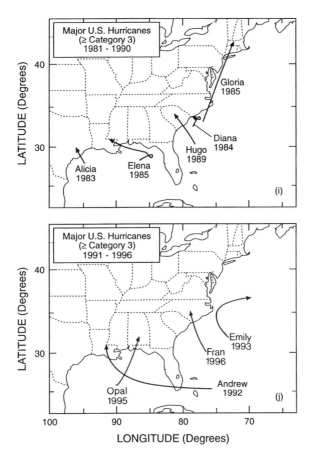

Figure C.5 (*continued*)

APPENDIX D

Selected Data and Names of Tropical Cyclones in Ocean Basins Around the World

This information was taken from the National Hurricane Center's website at: www.nhc.noaa.gov/names.html and from World Meteorological Organization, 1993: *Will There Be a Tropical Cyclone with Your Name?* Fact Sheet No. 11, May 1993.

Table D.1   Atlantic names – these lists are recycled every six years

1997	1998	1999	2000	2001	2002
Ana	Alex	Arlene	Alberto	Allison	Arthur
Bill	Bonnie	Bret	Beryl	Barry	Bertha
Claudette	Charley	Cindy	Chris	Chantal	Cesar
Danny	Danielle	Dennis	Debby	Dean	Dolly
Erika	Earl	Emily	Ernesto	Erin	Edouard
Fabian	Frances	Floyd	Florence	Felix	Fran
Grace	Georges	Gert	Gordon	Gabrielle	Gustav
Henri	Hermine	Harvey	Helene	Humberto	Hortense
Isabel	Ivan	Irene	Isaac	Iris	Isidore
Juan	Jeanne	Jose	Joyce	Jerry	Josephine
Kate	Karl	Katrina	Keith	Karen	Kyle
Larry	Lisa	Lenny	Leslie	Lorenzo	Lili
Mindy	Mitch	Maria	Michael	Michelle	Marco
Nicholas	Nicole	Nate	Nadine	Noel	Nana
Odette	Otto	Ophelia	Oscar	Olga	Omar
Peter	Paula	Philippe	Patty	Pablo	Paloma
Rose	Richard	Rita	Rafael	Rebekah	Rene
Sam	Shary	Stan	Sandy	Sebastien	Sally
Teresa	Tomas	Tammy	Tony	Tanya	Teddy
Victor	Virginie	Vince	Valerie	Van	Vicky
Wanda	Walter	Wilma	William	Wendy	Wilfred

Table D.2   Eastern North Pacific – these lists are recycled every six years

1997	1998	1999	2000	2001	2002
Andres	Agatha	Adrian	Aletta	Adolph	Alma
Blanca	Blas	Beatriz	Bud	Barbara	Boris
Carlos	Celia	Calvin	Carlotta	Cosme	Cristina
Dolores	Darby	Dora	Daniel	Dalila	Douglas
Enrique	Estelle	Eugene	Emilia	Erick	Elida
Felicia	Frank	Fernanda	Fabio	Flossie	Fausto
Guillermo	Georgette	Greg	Gilma	Gil	Genevieve
Hilda	Howard	Hilary	Hector	Henriette	Herman
Ignacio	Isis	Irwin	Ileana	Israel	Iselle
Jimena	Javier	Jova	John	Juliette	Julio
Kevin	Kay	Kenneth	Kristy	Kiko	Kenna
Linda	Lester	Lidia	Lane	Lorena	Lowell
Marty	Madeline	Max	Miriam	Manuel	Marie
Nora	Newton	Norma	Norman	Narda	Norbert
Olaf	Orlene	Otis	Olivia	Octave	Odile
Pauline	Paine	Pilar	Paul	Priscilla	Polo
Rick	Roslyn	Ramon	Rosa	Raymond	Rachel
Sandra	Seymour	Selma	Sergio	Sonia	Simon
Terry	Tina	Todd	Tara	Tico	Trudy
Vivian	Virgil	Vernoica	Vincente	Velma	Vance
Waldo	Winifred	Wiley	Willa	Wallis	Winnie
Xina	Xavier	Xina	Xavier	Xina	Xavier
York	Yolanda	York	Yolanda	York	Yolanda
Zelda	Zeke	Zelda	Zeke	Zelda	Zeke

Table D.3   Arafura Sea and Gulf of Carpentaria

Amelia	Alistair
Bruno	Bonnie
Coral	Craig
Dominic	Debbie
Esther	Evan
Ferdinand	Fay
Gretel	George
Hector	Helen
Irma	Ira
Jason	Jasmine
Kay	Kim
Laurence	Laura
Marian	Matt
Neville	Nicola
Olivia	Oswald
Phil	Penny
Rachel	Russell
Sid	Sandra
Thelma	Trevor
Vance	Valerie
Winsome	Warwick

Table D.4   Solomon Sea and Gulf of Papua

Adel
Epi
Guba
Ila
Kama
Tako
Upi

Table D.5   Northern Australian Region. Each region starts the year with whatever name is next on the list after the last cyclone of the previous year. When the end of a column is reached, the next name is the top name of the other column

Gretel	George
Hector	Helen
Irma	Ira
Jason	Jasmine
Kay	Kim
Laurence	Laura
Marian	Matt
Neville	Nicola
Olivia	Oswald
Phil	Penny
Rachel	Russell
Sid	Sandra
Thelma	Trevor
Vance	Valerie
Winsome	Warwick
Alistair	Amelia
Bonnie	Bruno
Craig	Coral
Debbie	Dominic
Evan	Esther
Fay	Ferdinand

Table D.6 Western Australian Region. Each region starts the year with whatever name is next on the list after the last cyclone of the previous year. When the end of a column is reached, the next name is the top name of the other column

Emma	Pedro
Frank	Rosita
Gertie	Sam
Hubert	Tina
Isobel	Vincent
Jacob	Walter
Kirsty	Alex
Lindsay	Bessi
Melanie	Chris
Nicholas	Daphne
Ophelia	Errol
Pancho	Fifi
Rhonda	Graham
Selwyn	Harriet
Tiffany	Ian
Victor	Jane
Alison	Ken
Billy	Lena
Connie	Monty
Damien	Naomi
Elsie	Oscar
Frederic	Pearl
Gwenda	Wuenton
Herbie	Sharon
Ilona	Tim
John	Vivienne
Kirrily	Willy
Leon	Annette
Marcia	Bobby
Ned	Chloe
Olga	Daryl

Table D.7 Eastern Australian Region. Each region starts the year with whatever name is next on the list after the last cyclone of the previous year. When the end of a column is reached, the next name is the top name of the other column

Monica	Agnes
Nigel	Barry
Odette	Celeste
Pierre	Dennis
Rebecca	Ethel
Sandy	Fergus
Tania	Gillian
Vernon	Harold
Wendy	Justin
Alfred	Katrina
Blanch	Les
Charlie	May
Denise	Nathan
Ernie	Olinda
Felicity	Pete
Greg	Rona
Hilda	Steve
Ivor	Tessi
Joy	Vaughan
Kelvin	Abigail
Lisa	Bernie
Mark	Claudia
Nina	Des
Oliver	Elinor
Polly	Fritz
Roger	Grace
Sadie	Harvey
Theodore	Ingrid
Violet	Jim
Warren	Kathy
	Lance

Table D.8 Fiji Region Names –Lists A, B, C, and D are used sequentially one after the other. The first name in any given year is the one immediately following the last name from the previous year. List E is a list of replacement names if they become necessary

List A	List B	List C	List D	List E
Ami	Arthur	Atu	Alan	Amos
Beni	Becky	Beti	Bart	Bobby
Cilla	Cliff	Cyril	Cora	Chris
Dovi	Daman	Drena	Dani	Daphne
Eseta	Elisa	Evan	Ella	Eva
Fili	Funa	Freda	Frank	Fanny
Gina	Gene	Gavin	Gita	Garry
Heta	Hettie	Hina	Hali	Helene
Ivy	Innis	Ian	Iris	Irene
Judy	Joni	June	Jo	Julie
Kerry	Kina	Keli	Kim	Ken
Lola	Lin	Lusi	Leo	Louise
Meena	Mick	Martin	Mona	Mike
Nancy	Nisha	Nute	Neil	Nat
Olaf	Oli	Osea	Oma	Odile
Percy	Prema	Pam	Paula	Pat
Rae	Rewa	Ron	Rita	Rene
Sina	Sarah	Susan	Sam	Sheila
Tam	Tomas	Tusi	Trina	Tui
Vaianu	Usha	Ursula	Uka	Ula
Wati	Vania	Veli	Vicky	Victor
Zita	William	Wini	Walter	Wilma
	Yasi	Yali	Yolande	Yalo
	Zaka	Zuman	Zoe	Zena

Table D.9 Southwest Indian Ocean – these lists are used sequentially, and they are not rotated every few years as are the Atlantic and Eastern Pacific lists

1997–1998	1998–1999	1999–2000	2000–2001	2001–2002	2002–2003	2003–2004	2004–2005
Aimay	Alda	Astride	Aviona	Alexina	Albertine	Agnielle	Antoinette
Bibianne	Birenda	Babiola	Babie	Bettina	Bentha	Bonita	Bordella
Cindy	Chikita	Connie	Colina	Cecilia	Christelle	Coryna	Chantelle
Donaline	Davina	Damienne	Dessilia	Daisy	Dorina	Doloresse	Daniella
Elsie	Ervina	Eline	Edwina	Edmea	Eliceca	Edwige	Elvina
Fiona	Francine	Felicia	Finella	Farah	Fodah	Flossy	Fabriola
Gemma	Genila	Gloria	Gracia	Geralda	Gail	Guylianne	Gretelle
Hillary	Helvetia	Hudah	Hutelle	Hollanda	Heida	Hansella	Helinda
Ireland	Irina	Innocente	Ionia	Ivy	Ingrid	Itelle	Iletta
Judith	Jocyntha	Jonna	Jourdanne	Julita	Josta	Jenna	Josie
Kimmy	Kristina	Kenetha	Konita	Kelvina	Kylie	Ketty	Karlette
Lynn	Lina	Lisanne	Laura	Litanne	Lidy	Lucia	Lisette
Monique	Marsia	Maizy	Monette	Mariola	Marlene	Molly	Maryse
Nicole	Naomie	Nella	Neige	Nadia	Natashia	Nadege	Nelda
Olivette	Orace	Ortensia	Octavie	Odile	Onelle	Odette	Ocline
Prisca	Patricia	Priscilla	Pamela	Pemma	Paulette	Paquerette	Phyllis
Renette	Rita	Rebecca	Rosita	Ronna	Raisa	Rolina	Rolina
Sarah	Shirley	Sophia	Stella	Sydna	Sabrina	Sylvianne	Sheryl
Tania	Tina	Terrence	Tasiana	Tella	Theresa	Talla	Thelma
Valencia	Vernoique	Victorine	Vigonia	Valentina	Vicky	Vivienne	Venyda
Wanicky	Wilvenia	Wilma	Wendy	Williana	Wilhelmine	Walya	Wiltina
Yandah	Yastride	Yuanselma	Yolande	Yvanna	Yvonne	Yoline	Yolette

Table D.10   Southeast Indian Ocean

Alex	Annette	Alison
Bessi	Bobby	Billy
Chris	Chloe	Connie
Daphne	Daryl	Damien
Errol	Emma	Elsie
Fifi	Frank	Frederic
Graham	Gertie	Gwenda
Harriet	Hubert	Herbie
Ian	Isobel	Ilona
Jane	Jacob	John
Ken	Kirsty	Kirrily
Lena	Lindsay	Leon
Monty	Margot	Marcia
Naomi	Nicholas	Ned
Oscar	Ophelia	Olga
Pearl	Pancho	Pedro
Quenton	Rhonda	Rosita
Sharon	Selwyn	Sam
Tim	Tiffany	Tina
Vivienne	Victor	Vincent
Willy		Walter

Table D.11   Western North Pacific Ocean – these names are used sequentially. This is a new name list used by the Joint Typhoon Warning center, effective at the start of 1996

Ann	Abel	Amber	Alex
Bart	Beth	Bing	Babs
Cam	Carlo	Cass	Chip
Dan	Dale	David	Dawn
Eve	Ernie	Ella	Elvis
Frankie	Fern	Fritz	Faith
Gloria	Greg	Ginger	Gil
Herb	Hannah	Hank	Hilda
Ian	Isa	Ivan	Iris
Joy	Jimmy	Joan	Jacob
Kirk	Kelly	Keith	Kate
Lisa	Levi	Linda	Leo
Marty	Marie	Mort	Maggie
Niki	Nestor	Nichole	Neil
Orson	Opal	Otto	Olga
Piper	Peter	Penny	Paul
Rick	Rosie	Rex	Rachel
Sally	Scott	Stella	Sam
Tom	Tina	Todd	Tanya
Violet	Victor	Vicki	Virgil
Willie	Winnie	Waldo	Wendy
Yates	Yule	Yanni	York
Zane	Zita	Zeb	Zia

218

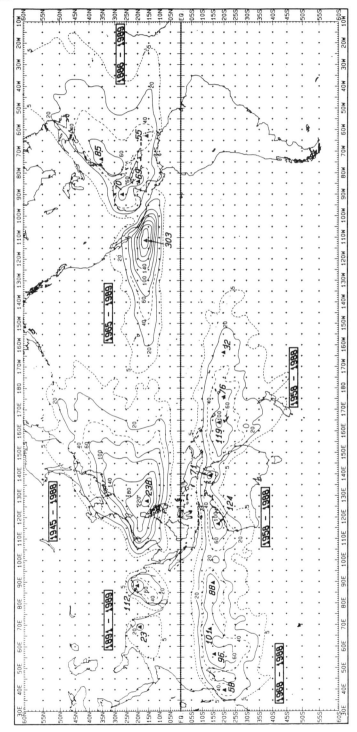

Figure D.1   Frequency of tropical cyclones per unit area. Contours show number of tropical cyclones per 100 years passing within 86 miles (139 kilometers) of any given point. Tropical cyclones which fail to reach at least tropical storm strength are not included. Solid triangles give locations of significant storm density maxima, while values of these maxima are given by adjacent bold print. Periods of record normalized to 100 years are shown for each ocean basin. Figure provided by C. Neumann, 1997; originally published by World Meteorological Organization (1993)

219

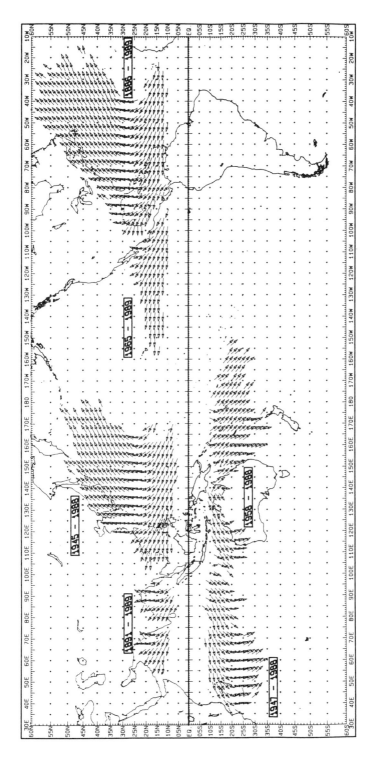

Figure D.2 Arrows depict mean direction of tropical cyclone motion averaged over specified periods of record. Tropical cyclones which failed to reach tropical storm strength are not included. Figure provided by C. Neumann, 1997; originally published by World Meteorological Organization (1993)

220   HURRICANES: THEIR NATURE AND IMPACT ON SOCIETY

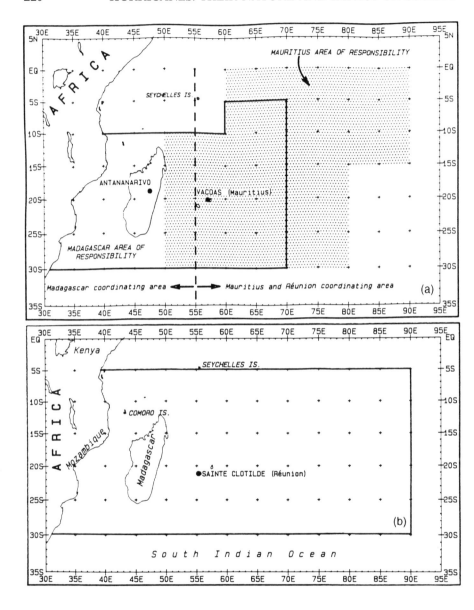

Figure D.3 (a–c) Regional Association I, high seas tropical cyclone warning responsibility areas for the southwest Indian Ocean. (a) Area for Madagascar (Antananarivo) and for Mauritius (Vacoas). (b) Area for Reunion (Sainte Clotilde). (c) Area for Mozambique (Maputo) and for Kenya (Nairobi)

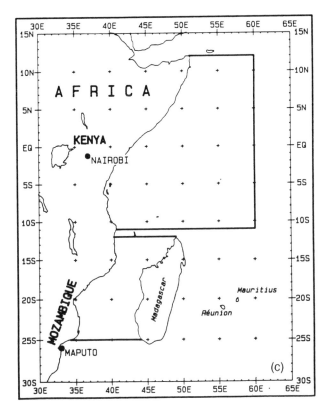

Figure D.3 (c) (*continued*)

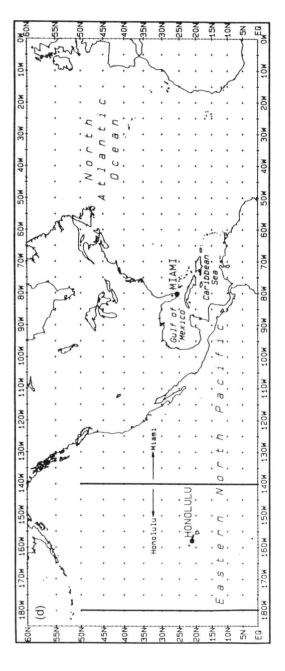

Figure D.3 (d)   Regional Association IV, United States National Hurricane Center, Miami, Florida area of high seas tropical cyclone warning responsibility. This includes (1) the North Atlantic, Caribbean Sea and the Gulf of Mexico and (2) the Eastern North Pacific east of longitude 140W. Also shown is the area of responsibility for the United States Central Pacific Hurricane Center located at Honolulu, Hawaii

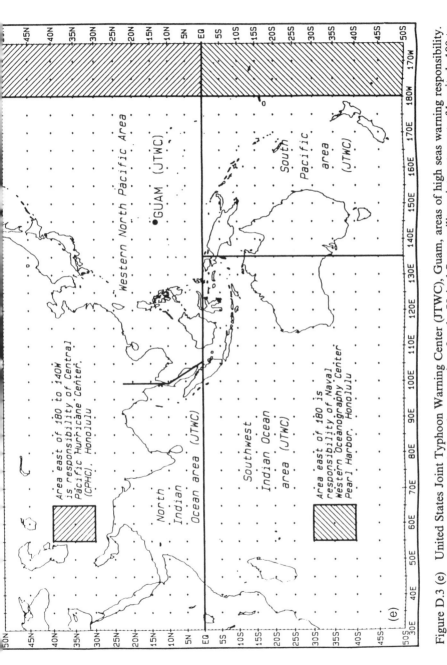

Figure D.3 (e)   United States Joint Typhoon Warning Center (JTWC), Guam, areas of high seas warning responsibility. JTWC is responsible for issuing tropical cyclone warnings for the United States military interests west of longitude 180 to the east coast of Africa

224

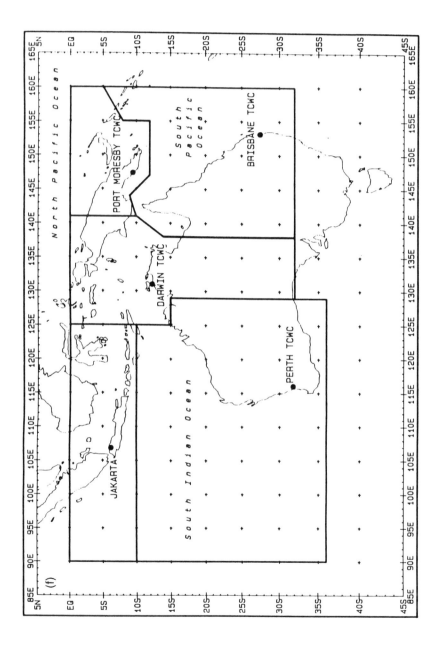

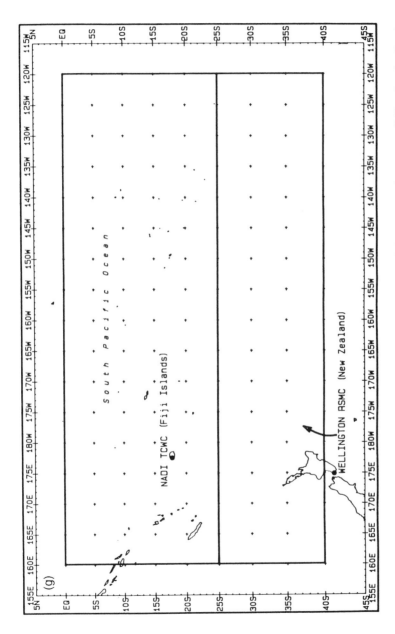

Figure D.3 (f, g) Regional Association V (WMO, 1989) areas of high seas warning responsibility for the South Pacific and the southeast Indian Ocean region. (f) Areas for Australia (Perth, Darwin and Brisbane), Papua New Guinea (Port Moresby) and Indonesia (Jakarta). (g) Areas for Fiji (Nadi) and New Zealand (Wellington)

226

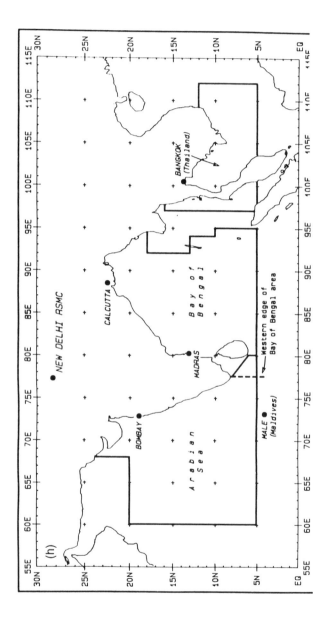

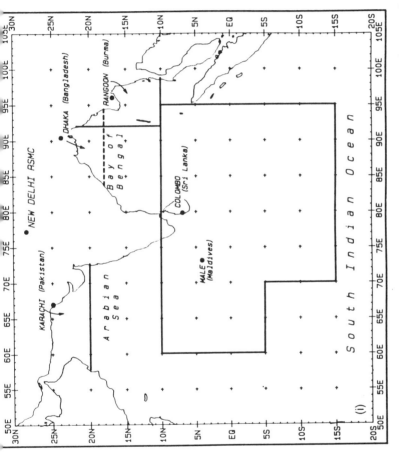

Figure D.3 (h, i) ESCAP/Panel on Tropical Cyclones (WMO, 1986), high seas tropical cyclone warning areas for the Bay of Bengal, the Arabian Sea and vicinity. (h) Areas of responsibility for India [Calcutta (Bay of Bengal) and Bombay (Arabian Sea)]. Calcutta bulletins for Bay of Bengal are re-broadcast by Madras. Also shown is Thailand (Bangkok) area of responsibility, eastern portion of which is in Typhoon Committee Region. (i) Areas of responsibility for Pakistan (Karachi), Bangladesh (Dhaka), Burma (Rangoon) and Sri Lanka (Columbo)

228

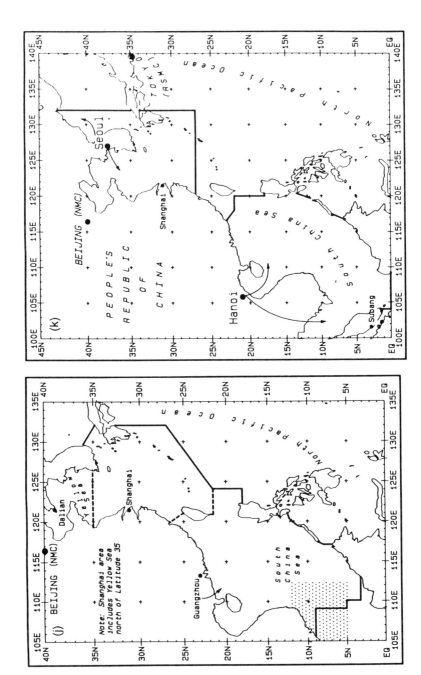

229

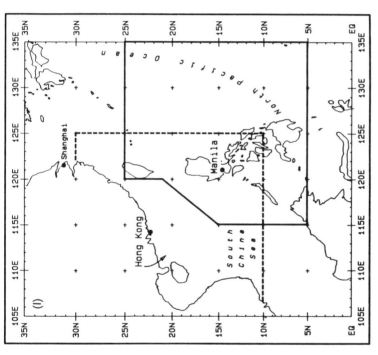

Figure D.3 (j–l) ESCAP/Typhoon Committee (WMO, 1987), high seas tropical cyclone warning areas for western North Pacific. (j) Areas of responsibility for People's Republic of China (Dalian, Shanghai and Guangzhau). Also shown (dot pattern) is Easternmost edge of Thailand area. See (h) for complete depiction of Thailand area of high seas forecast responsibility. (k) Areas of responsibility for Republic of Korea (Seoul), Viet Nam (Hanoi) and Japan (Tokyo). The Japanese area of high seas responsibility is not specifically shown but includes entire western North Pacific basin. (l) Areas of responsibility for Philippine Islands (Manila, PAGASA) and Hong Kong (Royal Observatory)

(a)

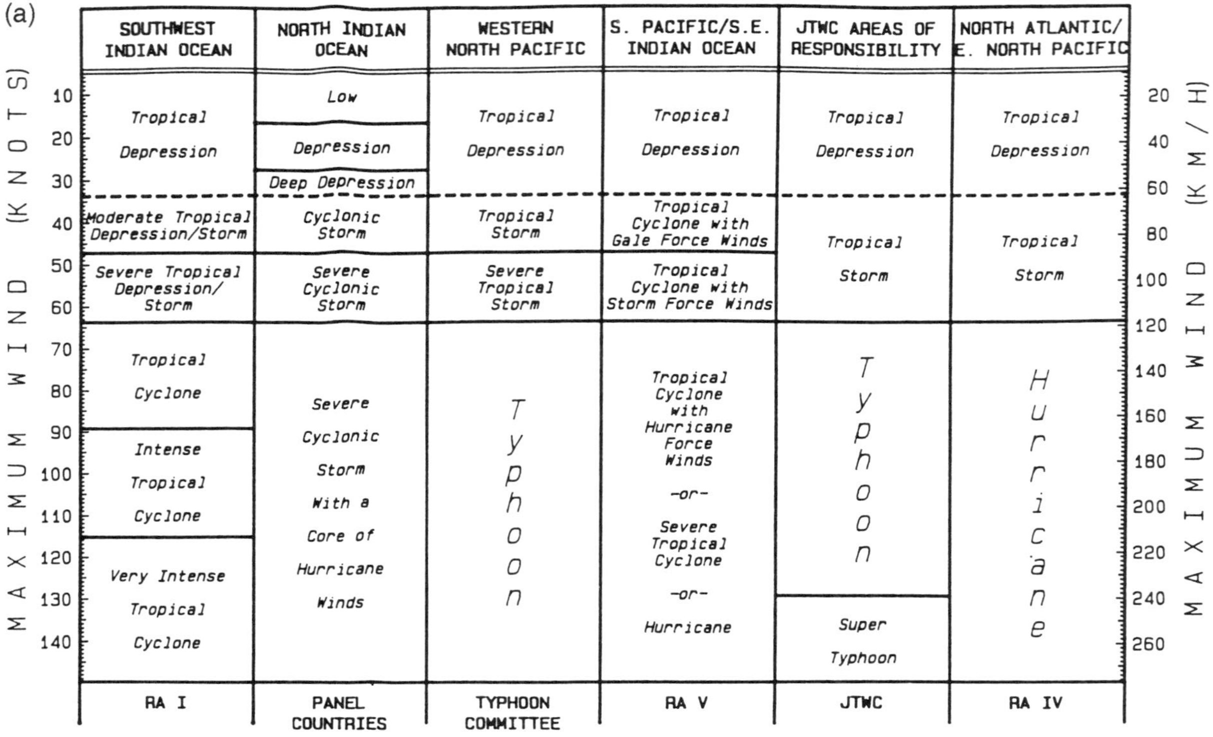

MAXIMUM WIND (KM/H)

20 40 60 80 100 120 140 160 180 200 220 240 260

MAXIMUM WIND (KNOTS)

10 20 30 40 50 60 70 80 90 100 110 120 130 140

SOUTHWEST INDIAN OCEAN	NORTH INDIAN OCEAN	WESTERN NORTH PACIFIC	S. PACIFIC/S.E. INDIAN OCEAN	JTWC AREAS OF RESPONSIBILITY	NORTH ATLANTIC/ E. NORTH PACIFIC
Tropical Depression	Low / Depression / Deep Depression	Tropical Depression	Tropical Depression	Tropical Depression	Tropical Depression
Moderate Tropical Depression/Storm		Tropical Storm	Tropical Cyclone with Gale Force Winds	Tropical Storm	Tropical Storm
Severe Tropical Depression/Storm	Cyclonic Storm	Severe Tropical Storm	Tropical Cyclone with Storm Force Winds	Typhoon	Hurricane
Tropical Cyclone	Severe Cyclonic Storm	Typhoon	Tropical Cyclone with Hurricane Force Winds -or- Severe Tropical Cyclone -or- Hurricane		
Intense Tropical Cyclone	Severe Cyclonic Storm With a Core of Hurricane Winds				
Very Intense Tropical Cyclone				Super Typhoon	
RA I	PANEL COUNTRIES	TYPHOON COMMITTEE	RA V	JTWC	RA IV

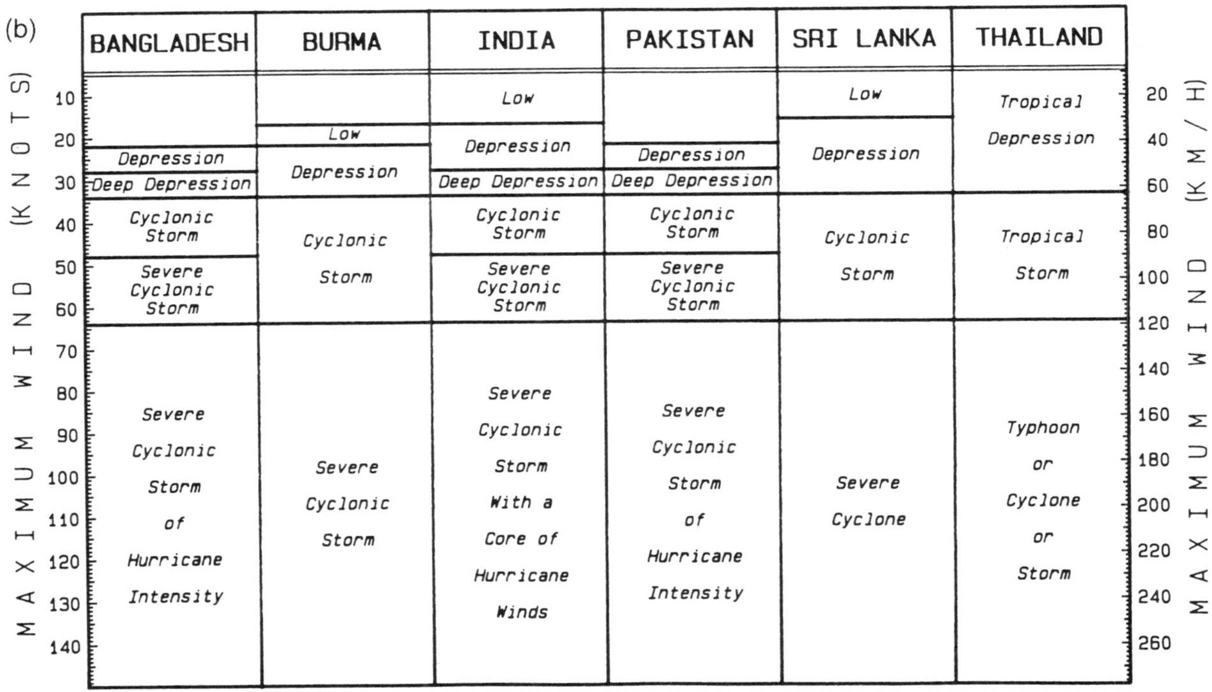

Figure D.4 (a) Classification of tropical cyclones by regional bodies. Definitions for North India Ocean vary by country; example shown is for India. Dashed line gives threshold storm intensity for naming or numbering of tropical cyclones. JTWC entries apply to their Western N. Pacific area of forecast responsibility. (b) Classification of cyclonic disturbances over North Indian Ocean area (WMO, 1986)

MAXIMUM WIND (KNOTS)	MAXIMUM WIND (KM/H)	BANGLADESH	BURMA	INDIA	PAKISTAN	SRI LANKA	THAILAND
10	20			Low		Low	Tropical Depression
20	40	Depression	Low	Depression	Depression	Depression	
30	60	Deep Depression	Depression	Deep Depression	Deep Depression		
40	80	Cyclonic Storm	Cyclonic Storm	Cyclonic Storm	Cyclonic Storm	Cyclonic Storm	Tropical Storm
50	100	Severe Cyclonic Storm		Severe Cyclonic Storm	Severe Cyclonic Storm		
60	120		Severe Cyclonic Storm				
70	140	Severe Cyclonic Storm of Hurricane Intensity		Severe Cyclonic Storm With a Core of Hurricane Winds	Severe Cyclonic Storm of Hurricane Intensity	Severe Cyclone	Typhoon or Cyclone or Storm
80	160						
90	180						
100	200						
110	220						
120	240						
130	260						
140							

APPENDIX E

Guide for Local Hurricane Decision-Makers

TIME DELINEATING SCHEDULE (TDS) FOR STORM EMERGENCIES

PREPARED BY:

LEE COUNTY EMERGENCY MANAGEMENT
P.O. BOX 398
2665 ORTIZ AVENUE
FORT MYERS, FLORIDA 33902-0398

July 1992

TIME DELINEATING SCHEDULE (TDS)
An Overview

Government plays a major role in protecting life and property from both natural and technological hazards by developing emergency operation plans to guide a community's response to and recovery from disasters or emergencies. But just as a comprehensive plan requires a zoning ordinance and/or a development standards ordinance to implement policies governing a community's growth and development, so too does an emergency plan require a tool to steer the complicated decision-making process through a crisis. Such a tool must not only clarify what actions should be taken and when they should be taken, but also account for the uncertainty present under any emergency which is caused by the particular characteristics of the threatening hazards.

The tool which enables decision makers to implement the emergency planning effort in Lee County is called the Time Delineating Schedule or TDS. Developed by Lee County, TDS provides a step-by-step process to trigger actions by decision-makers in preparation for, response to, and if required, recovery from an emergency or disaster. It accomplishes this by defining ten distinct time periods that describe key phrases or objectives of the emergency. These phases, which are presented in the flow chart and an explanation on the preceding pages, describe a logical sequence to follow for implementing actions according to prescribed needs and priorities.

Specific actions are assigned to each phase that are designed not only to meet the intended objectives for that phase, but also to lay the foundation for the next set of actions required to meet the objectives of the following phase. Decision-makers review, analyze and implement those actions that must, or can be taken based on the threatening hazard's extent and magnitude, and on such constraints as the hazard's speed of onset. Because each phase and its actions serve as a building block for succeeding phases, TDS provides the decision-maker with a timetable for completing actions, while reducing the possibility of not implementing an action that may delay or hinder another action from taking place later on in the emergency.

The TDS concept consists of a manual describing the time delineating process, recommended actions for each phase, and a status of action checklist. Although originally designed for the hurricane threat, TDS can be used in other emergency situations by modifying the time frame for each phase according to the hazard's characteristics and length of the warning period.

A-1

In sum, TDS provides an umbrella for defining, guiding and documenting decision-makers actions through the various phases of an emergency. Its primary aim is to ensure that people and property are properly protected in time and that the necessary human and physical resources are in place to support such protective actions. It can be modified to account for the constraint of the hazard itself, the response time available, and other conditions which define the uncertainty of the decision-making environment. Finally, it is used to inform the public on actions the County has taken to reduce the exposure risk to hazardous effects.

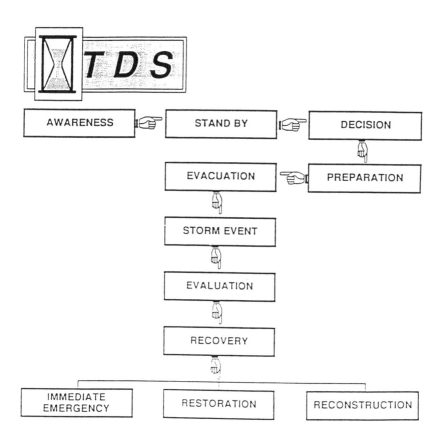

A-2

PHASE	DESCRIPTION
☐ AWARENESS	A period of time, usually consisting of twelve (12) hours commencing at seventy-two (72) hours to approximately sixty (60) hours before extrapolated landfall. This is the notification period, during which appropriate agencies and organizations (public, quasi-public, and private) should be made aware of the situation.
☐ STAND-BY	A period of time, usually consisting of ten (10) hours commencing at sixty (60) hours to approximately forty-eight (48) hours before extrapolated landfall. This is the alert period for the acceleration of preparedness actions for emergency and vital services affected by the situation.
☐ DECISION	A period of time, usually consisting of three (3) hours commencing at forty-eight (48) hours to approximately thirty-six (36) hours before extrapolated landfall. During this period, the decision to evacuate must be reached and the possibility of the evacuation order be made public. This is the period during which the populace should take precautionary actions in order to cope with the threatening situation.
☐ PREPARATION	A period of time, usually consisting of nine (9) hours commencing at, forty-five (45) hours to approximately thirty-three (33) hours before extrapolated landfall. This is the re-analysis period, and the preparation time needed to place emergency personnel and resources into position for operations.

A-3

☐ EVACUATION

A period of time commencing at that point (36 to 24 hours before extrapolated landfall) when Lee County officials determine and announce the official evacuation order, continuing until that point either prior to the estimated time of sustained tropical storm force winds (39 miles per hour), or prior to the estimated time of inundation (one-foot) of evacuation routes caused by either the storm surge or fresh water flooding. This is the commencement through completion of the relocation period; all evacution activities must be completed.

☐ STORM EVENT

A period of time, commencing with the arrival of sustained tropical storm force winds (39 miles per hour), or the inundation of primary evacuation routes, continuing until that point when the local government determines and issues the "ALL CLEAR" announcement. This is the in-place shelter period for the threatened populace, either sheltered in private homes or designated public buildings throughout the County.

☐ EVALUATION

A period of time, consisting of several days to a couple of weeks commencing at that point when sustained winds decrease to forty miles per hour (40 mph) or below. This is the evaluation and assessment period, where Lee County officials intitially assess and prioritize the emergency situation and/or generate requirements.

☐ IMMEDIATE
EMERGENCY

A period of time, lasting from a couple of weeks up to several months after the storm event. This is the first phase of the recovery period where Lee County public safety agencies and non-governmental organizations respond and provide immediate emergency assistance to prioritized requirements.

☐ RESTORATION

A period of time, consisting of several months to a couple of years after the storm event. This is the second phase of the recovery period where Lee County officials coordinate the repair of the public infrastructure and primarily focus on social and economic activities that will return the community to pre-storm levels.

☐ RECONSTRUCTION

A period of time, consisting of a couple of years to several years after the storm event. This is the last and longest phase of the recovery period where Lee County officials will focus on activities that will mitigate future storm damages.

PHASE	RESPONSE ACTIONS	NUMBER OF RESPONSE ACTIONS
Awareness	1-32	32
Stand-By	33-69	37
Decision	70-92	23
Preparation	93-106	14
Evacuation	107-112	8
Storm Event	115-117	3
Evaluation	118-127	10
Immediate Emergency	128-145	18
Restoration	146-171	26
Reconstruction	172-176	5

A-5

TIME DELINEATING SCHEDULE (TDS)

IF SITUATION   * * * * * * * * 6 RESPONSE ACTIONS * * * * * * * * * *
WARRANTS
P (a/p)   ☐   1.   Monitor hazardous weather conditions in the
                           Atlantic Ocean, Caribbean Sea, and Gulf of
                           Mexico.

P (a/p)   ☐   2.   Coordinate with the National Weather Service
                           (NWS) concerning meterological information
                           availability.

P (a/p)   ☐   3.   Coordinate with the County's consultant for
                           meterological services.

P (a/s)   ☐   4.   Coordinate with officials from the State Division
                           of Emergency Management, local municipalities,
                           surrounding counties, and other emergency-related
                           officials.

P (a/s)   ☐   5.   Compile and transmit the GDS report (storm
                           forecast) to appropriate agencies, organizations
                           and groups via facsimile machine.

T (a/p)   ☐   6.   Disseminate hurricane preparedness information
                           via the broadcast and print media outlets.

AWARENESS   * * * * * * * * 26 RESPONSE ACTIONS * * * * * * * * *

P (a/o)   ☐   7.   Activate the Lee County Emergency Operations
                           Center (EOC) with essential personnel. Review
                           assignments with County DEM staff.

P (a/p)   ☐   8.   Activate storm tracking and assessment system.

P (a/s)   ☐   9.   Establish liaison with appropriate governmental
                           and non-governmental emergency-related officials,
                           agencies and organizations.

P (p)   ☐   10.   Coordinate and disseminate all County public
                           information activities.

P (p)   ☐   11.   Perform a hurricane vulnerability analysis of the
                           threatening storm emergency and revise as
                           situation warrants.

T (s)   ☐   12.   Establish and maintain a log of events and/or
                           actions.

EM ASSIGNMENT: (a) = Administration (p) = Planning (o) = Operations
               (r) = Resource Management (s) = Staff
LEVEL OF PRIORITY: P = Primary S = Secondary T = Tertiary

A-6

S (a/s) ☐ 13. Alert and brief County Commissioners, appropriate
administrative staff members and local munici-
palities on the threatening storm emergency.

P (o) ☐ 14. Prepare .Lee County EOC according to floor plan
under emergency conditions.

P (o) ☐ 15. Acquire extra telephones and facsimile machines.
Test all EOC telephone equipment.

P (o) ☐ 16. Activate storm messages on hold button of EOC
telephone system.

T (o/p) ☐ 17. Coordinate the proper placement of evacuation
signage, as applicable

P (p) ☐ 18. Activate the Phone Notification System (PNS), as
applicable.

P (p) ☐ 19. Issue storm information reports via the fax
machine.

P (p) ☐ 20. Activate the Emergency Hotline capability of the
information network.

S (o/p) ☐ 21. Begin exchanging meterological information with
the SWFR Airport Operations staff.

S (o/p) ☐ 22. Prepare for the utilization of primary evacuation
routes - make temporary repairs to existing road
construction projects or prepare to delay start
of any new projects.

S (a/s) ☐ 23. Request all County Department Directors to
designate their personnel as essential and non-
essential according to their storm emergency-
related responsibilities or assignments.

S (a/s) ☐ 24. Request all Department Directors to review and/or
implement emergency plans for the protection of
County facilities and equipment.

S (a) ☐ 25. Request Department Directors to cancel all leaves
for County personnel.

P (o) ☐ 26. Test EOC communications equipment.

P (o) ☐ 27. Top off fuel tanks of emergency generators at EOC
and monitor.

EM ASSIGNMENT: (a) = Administration (p) = Planning (o) = Operations
                (r) = Resource Management (s) = Staff
LEVEL OF PRIORITY: P = Primary S = Secondary T = Tertiary

P (o) ☐ 29. Test EOC emergency utility system (i.e., electri-
                city, water, and sewer). Arrange for additional
                toilet provisions.

S (a/p) ☐ 30. Issue   public   information   statements,   as
                applicable.

T (o/p) ☐ 31. Report actual tide and wind measurements to the
                National Weather Service (NWS).

T (o) ☐ 31. Check the operability of the NOAA Weather Alert
                Radio transmitter and monitor.

S (p) ☐ 32. Establish initial coordination with the assigned
                manager and coordinator's of Lee County's Storm
                Information Hotline (SIHL) Center   and   brief
                staff.

STAND-BY   * * * * * * * 37 RESPONSE ACTIONS * * * * * * * * *

S (a/p) ☐ 33. Activate the County public information officer
                (PIO).

P (r) ☐ 34. Fuel all County vehicles and essential equipment
                to capacity.

P (o) ☐ 35. Issue access clearance badges to EOC officials.

P (o/p) ☐ 36. Establish emergency information phones in EOC and
                brief staff.

S (p) ☐ 37. Notify E-911 answering points that the SIHL has
                been activated.

T (o) ☐ 38. Notify EOC radio repair company of threatening
                storm emergency and potential service require-
                ments.

S (a/p) ☐ 39. Issue public information statements, as
                applicable.

T (r) ☐ 40. Correct any deficiencies found in County facili-
                ties, vehicles and equipment   utilized for
                emergency activities.

T (o) ☐ 41. Secure a Lee County crane (or aerial ladder) to
                be on standby outside the EOC for communications
                tower emergency needs.

EM ASSIGNMENT: (a) = Administration (p) = Planning (o) = Operations
               (r) = Resource Management (s) = Staff
LEVEL OF PRIORITY: P = Primary S = Secondary T = Tertiary

S (o)   ☐   42. Arrange for parking, feeding and sleeping arrangements (including transportation) for EOC occupants.

P (r)   ☐   43. Make arrangements with Lee County Fleet Management to have garage and depots operate on a 24 hour basis when the preparation phase begins.

S (a)   ☐   44. Request that all County Department Directors brief employees of emergency responsibilities for both pre-storm and post-storm operations.

P (o)   ☐   45. Implement interior and exterior security systems and plans for EOC.

P (o/p) ☐   46. Commence coordination of the traffic movement plan (i.e., control points & devices).

P (a/r) ☐   47. Make arrangements with companies to have port-a-johns delivered to shelter locations. Coordinate with Public Health, School Board and American Red Cross officials.

T (o/p) ☐   48. Report actual tide and wind measurements to the National Weather Service.

P (a/p/r) ☐ 49. Commence coordination of the emergency public sheltering plan (i.e., designation, staffing and supplies). Activate Lee County School District, American Red Cross and State HRS officials.

P (r)   ☐   50. Notify the Lee County Health Department to commence acquisition of nurses, doctors, portable toilets, and other supplies to support shelter operations.

P (o)   ☐   51. Secure space (42 cubic feet) in the EOC for refrigerator and freezer storage to accommodate Lee County Public Health vaccines.

S (r)   ☐   52. Notify the Lee County Humane Society to initiate emergency procedures for the support of the pet/animal shelter.

T (o/p) ☐   53. Monitor traffic conditions.

T (o)   ☐   54. Acquire a backup duplicating machine for the EOC.

P (a/p) ☐   55. Activate the Emergency Broadcast System (EBS).

EM ASSIGNMENT: (a) = Administration (p) = Planning (o) = Operations
               (r) = Resource Management (s) = Staff
LEVEL OF PRIORITY: P = Primary S = Secondary T = Tertiary

A-9

P (o) ☐ 56. Notify the RACES group of the threatening storm emergency.

P (a/o) ☐ 57. Restrict the general public from entrance into the EOC.

P (a/p) ☐ 58. Restrict recreational vehicles, trailered boats or campers to Sanibel and Captiva Islands.

S (a/p) ☐ 59. Restrict visitors to Sanibel and Captiva Islands.

T (a/p) ☐ 60. Advise the movement of all slow moving vehicles (less than 25 mph) from barrier islands and low-lying areas.

S (a/p) ☐ 61. Advise boat owners to secure and prepare their property for severe weather conditions and for a possible marine evacuation of the coastal waters.

S (a/p) ☐ 62. Advise island residents to secure their property for severe weather conditions and for a possible boat evacuation off coastal islands.

S (a/p) ☐ 63. Advise construction companies to secure all construction sites of materials or equipment against displacement by wind forces.

P (a/r/p) ☐ 64. Request construction companies to remove port-a-johns from job sites or deliver to specific location to assist shelter operations.

T (a/p) ☐ 65. Advise area businesses to secure their property against displacement by wind forces.

S (a/p) ☐ 66. Advise beach motel/hotel businesses of the potential storm emergency - evacuation may be required.

P (a/o/p) ☐ 67. Advise and coordinate operations of drawbridges throughout the County.

P (r) ☐ 68. Top off all County fuel dispensing tanks and position emergency power generators at locations.

P (r) ☐ 69. Coordinate the establishment of an emergency worker shelter (i.e., designation, staffing, and supplies).

EM ASSIGNMENT: (a) = Administration (p) = Planning (o) = Operations (r) = Resource Management (s) = Staff
LEVEL OF PRIORITY: P = Primary S = Secondary T = Tertiary

DECISION   * * * * * * * * * * 23 ACTIONS * * * * * * * * * * *

P (a/p) ☐ 70. Advise EBS primary control station to relocate and operate out of EOC.

P (a)   ☐ 71. Coordinate with County constitutional officers on either closing or limiting County business and/or services.

P (a)   ☐ 72. Recommend or advise the Lee County School Board to close schools.

P (a)   ☐ 73. Recommend or advise private schools to close.

P (a/p) ☐ 74. Transfer the SIHL to another designated Lee County office and brief staff.

P (a)   ☐ 75. Brief County Commissioners on the threatening storm emergency.

S (a/p) ☐ 76. Advise early recommended evacuation of the barrier islands and low-lying areas - no emergency public shelters will be open.

S (a/p/o) ☐ 77. Advise and coordinate a marine evacuation of the coastal waters.

S (a/p) ☐ 78. Advise and coordinate evacuation of off-shore islands utilizing boats.

P (a/p/r) ☐ 79. Activate shelter managers and officials to pre-determined locations.

P (a/o) ☐ 80. Activate RACES members to pre-determined locations.

S (r)   ☐ 81. Notify the Humane Society (Animal Control) to be prepared to pick-up animals at emergency public shelters, as necessary.

S (a)   ☐ 82. Advise cancellation of public social events.

S (a/p) ☐ 83. Issue public information statements, as necessary.

T (o/p) ☐ 84. Report actual tide and wind measurements to the National Weather Service.

T (o/p) ☐ 85. Evaluate observed traffic situations and correct deficiencies.

EM ASSIGNMENT: (a) = Administration (p) = Planning (o) = Operations
                (r) = Resource Management (s) = Staff
LEVEL OF PRIORITY: P = Primary S = Secondary T = Tertiary

P (o/p) ☐ 86. Activate the traffic control plan:

   P ☐ Traffic Control Points

   S ☐ Traffic Control Devices

P (r) ☐ 87. Coordinate emergency transportation requirements (i.e., vehicles, drivers, verification of people with special needs, and the designation of pick-up points).

S (a/p) ☐ 88. Advise against visiting the islands.

P (o/r) ☐ 89. Relocate essential emergency equipment and vehicles to pre-determined locations.

S (r) ☐ 90. Notify tow-truck businesses of the potential storm emergency and pre-determine wrecker locations along critical evacuation routes.

P (a) ☐ 91. Advise the Chairperson of the Board of County Commissioners to declare a state of local emergency for Lee County.

P (a/s) ☐ 92. Coordinate and advise state of local emergency with the following:

   P ☐ State Division of Emergency Management (DEM)

   P ☐ National Hurricane Center (NHC)

   P ☐ City of Cape Coral

   P ☐ City of Fort Myers

   P ☐ City of Sanibel

   S ☐ Charlotte County

   S ☐ Collier County

   S ☐ Hendry County

   T ☐ Glades County

   T ☐ Sarasota County

EM ASSIGNMENT: (a) = Administration (p) = Planning (o) = Operations
(r) = Resource Management (s) = Staff
LEVEL OF PRIORITY: P = Primary S = Secondary T = Tertiary

A-12

PREPARATION		* * * * * * * * 14 RESPONSE ACTIONS * * * * * * * * *
P (o)	☐	93. Establish and affirm communications with shelter and/or deployed emergency personnel.
T (a/p)	☐	94. Report tide and wind measurements to the National Weather Service.
P (a/p)	☐	95. Issue public information statements.
P (a/p)	☐	96. Disseminate emergency information, advisories and bulletins via the facsimile machine to surrounding counties, State DEM and other emergency-related agencies or organizations.
P (a/r)	☐	97. Implement 24-hour operation of Fleet Management garage and fueling depots.
S (p/o)	☐	98. Evaluate traffic situations and correct deficiencies.
S (a/p/o)	☐	99. If determined applicable, restrict all traffic seeking access to Sanibel and Captiva Islands at intermittent periods to allow two and possibly three lanes to exit the Island.
S (a/p/o)	☐	100. If determined applicable, restrict all traffic seeking access to Pine Island/Matlacha at intermittent periods to allow two lanes to exit the Island.
S (a/p/o)	☐	101. If determined applicable, recommend Charlotte County to restrict all traffic seeking access to Gasparilla Island at intermittent periods to allow two lanes to exit the Island.
P (a/p/o)	☐	102. Coordinate with the State Division of Emergency Management (DEM) concerning the following items:
P	☐	When the evacuation order will be issued by the State and the County.
S	☐	Estimation of population-at-risk.
S	☐	Number of shelters required inland.
P	☐	State assistance needs:
P	☐	Law enforcement personnel
P	☐	Traffic control

EM ASSIGNMENT: (a) = Administration (p) = Planning (o) = Operations
(r) = Resource Management (s) = Staff
LEVEL OF PRIORITY: P = Primary S = Secondary T = Tertiary

P   ☐   Security

    P   ☐   Shelter personnel

P   ☐   Accessibility of evacuation routes.

P   ☐   Need for Governor to issue an executive
                    order to support County operations.

P (a/p/o)   ☐   103. Inform the State Division of Emergency Management
                     (DEM) of the following protection actions:

P   ☐   Evacuation

S   ☐   Public Sheltering

P   ☐   Road/Bridge Closures

P (a/p)   ☐   104. Advise recommended evacuation of following resi-
                     dents:

P   ☐   People with Special Needs

P   ☐   People without Transportation

P   ☐   Islands

P   ☐   Low-Lying Areas

S   ☐   Tourists

S   ☐   Mobile Homes

S   ☐   Manufactured Housing

S   ☐   Recreational Vehicles (RV's)

S   ☐   Campers

P (a)   ☐   105. Advise and coordinate a recommended evacuation
                     with surrounding counties.

P (a)   ☐   106. Activate emergency transportation resources.

EM ASSIGNMENT: (a) = Administration (p) = Planning (o) = Operations
               (r) = Resource Management (s) = Staff
LEVEL OF PRIORITY: P = Primary   S = Secondary   T = Tertiary

EVACUATION   * * * * * * * * 8 RESPONSE ACTIONS * * * * * * * * *

P (a)   ☐   107. Advise the Chairperson of the Board of County Commissioners to issue an evacuation order for areas vulnerable to life-threatening conditions.

P (p/r)   ☐   108. Continue phasing - of emergency public shelter openings and placement of shelter signs.

P (o)   ☐   109. Maintain emergency public shelter communications.

P (a/s)   ☐   110. Monitor emergency public shelter conditions and correct deficiencies.

P (o)   ☐   111. Activate EOC emergency utility systems.

S (a/o)   ☐   112. Advise and coordinate the shut-down of public and private utility systems.

P (p/o)   ☐   113. Observe traffic situations and correct deficiencies.

P (a/s)   ☐   114. Commence coordination of post-storm response planning activities:

   S   ☐   Search & Rescue

   P   ☐   Emergency Medical Care

   S   ☐   Care of Dead

   S   ☐   Security Check Points

   P   ☐   Return of Evacuees

   S   ☐   Emergency Regulations

   S   ☐   Preliminary Damage Assessment

   S   ☐   Portage Areas

   P   ☐   Procurement of Supplies

   S   ☐   Public Health Monitoring

   S   ☐   Assessment of Community Needs

   S   ☐   Emergency Relief Assistance

   S   ☐   Restoration of Critical Lifelines

EM ASSIGNMENT: (a) = Administration (p) = Planning (o) = Operations
                (r) = Resource Management (s) = Staff
LEVEL OF PRIORITY: P = Primary   S = Secondary   T = Tertiary

A-15

S	☐	Removal of Debris
S	☐	Emergency Worker Stations
S	☐	Recovery Centers
S	☐	Building Moratoriums
S	☐	Recovery Task Force
S	☐	Staging Areas
S	☐	Emergency Distribution Centers
T	☐	Federal Public Assistance
T	☐	Disaster Field Offices (DFOs)
T	☐	Disaster Application Centers (DACs)
T	☐	Presidential Declaration
T	☐	Temporary Housing
T	☐	EOC De-Briefing

STORM EVENT   * * * * * * * * * 3 RESPONSE ACTIONS * * * * * * * * *

P (a/p) ☐ 115. Monitor storm characteristics.

P (o/p) ☐ 116. Continue emergency public shelter communications.

P (a/s) ☐ 117. Continue post-storm response planning activities.

EVALUATION   * * * * * * * * * 10 RESPONSE ACTIONS * * * * * * * *

P (a/p) ☐ 118. Determine if the primary threat still exists from appropriate agencies.

P (o/p) ☐ 119. Conduct and coordinate the initial emergency assessment of situation.

S (a/p) ☐ 120. Determine and prioritize emergency-generated requirements.

P (a/s) ☐ 121. Re-establish and affirm communications with the following:

EM ASSIGNMENT: (a) = Administration (p) = Planning (o) = Operations
               (r) = Resource Management (s) = Staff
LEVEL OF PRIORITY: P = Primary S = Secondary T = Tertiary

P　　　☐　Emergency Public Shelters

P　　　☐　Deployed Emergency Personnel

P　　　☐　State Division of Emergency Management (DEM)

P　　　☐　City of Cape Coral

P　　　☐　City of Fort Myers

P　　　☐　City of Sanibel

S　　　☐　Charlotte County

S　　　☐　Collier County

T　　　☐　Hendry County

T　　　☐　Glades County

P (a/s)　☐　122. Re-mobilize emergency operational agencies, organizations and private resources.

S (a)　☐　123. Enact emergency resolutions, ordinances, suspensions of administrative rules and/or procedures.

P (a/s)　☐　124. Activate the Lee County Disaster Purchase Order System if the county's automated purchasing and procurement program is rendered inoperative.

S (a/s)　☐　125. Complete and transmit a Emergency Incident Report to the State Division of Emergency Management (DEM).

P　　　☐　126. Commence clearance of the runways of the SW Florida Regional and Lee County Airports.

S (a/p/o)　☐　127. If State damage assessment assistance is required:

P　　　☐　Appoint County/City personnel as guides.

S　　　☐　Arrange for transportation.

S　　　☐　Obtain maps of areas to be surveyed.

EM ASSIGNMENT: (a) = Administration (p) = Planning (o) = Operations
　　　　　　　 (r) = Resource Management (s) = Staff
LEVEL OF PRIORITY:　P = Primary　S = Secondary　T = Tertiary

A-17

IMMEDIATE   * * * * * * * * 18 RESPONSE ACTIONS * * * * * * * *
EMERGENCY

P (a/s) ☐   128. Commence local emergency response to prioritize
                    generated requirements.

P (a/s) ☐   129. Activate appropriate response plans:

   P   ☐   Care of the Injured and/or Dead

   P   ☐   Security Check Points

   S   ☐   Request Relief Assistance

      S   ☐   Food

      P   ☐   Water/Ice

      T   ☐   Clothing

      P   ☐   Shelter

      T   ☐   Crisis Counseling

      T   ☐   Emergency Loans/Grants

   P   ☐   Restoration of Critical Lifelines:

      P   ☐   Electricity

      P   ☐   Water

      S   ☐   Transportation:

         S   ☐ Air

         S   ☐ Land

         T   ☐ Water

      P   ☐   Communications

P (a/p) ☐   130. Issue the "ALL CLEAR" announcement for designated
                    areas.

P (a/p) ☐   131. Activate and mobilize the recovery task force and
                    perform the following:

   P   ☐   Review   damage   reports   and   identify
                     mitigation opportunities.

EM ASSIGNMENT: (a) = Administration (p) = Planning (o) = Operations
               (r) = Resource Management (s) = Staff
LEVEL OF PRIORITY: P = Primary S = Secondary T = Tertiary

P   ☐ Recommend   emergency   resolutions   and
              ordinances   pertaining   to post-hurricane
              activities.

S   ☐ Recommend   changes   to   land   development
              regulations.

P   ☐ Formulate recommendations to guide community
              recovery.

S   ☐ Formulate   special   committees   and   sub-
              committees to complete specific tasks.

P   ☐ Initiate   hazard   mitigation   projects   and
              programs for state or federal funding.

P   ☐ Participate   in   state   and   federal   hazard
              mitigation efforts.

T   ☐ Review   emergency   actions   and   recommend
              amendments   to   emergency   plans   and
              procedures.

S   ☐ Appoint   or   acquire   disaster   recovery
              coordinator.

S   ☐ Appoint   or   acquire   economic   recovery
              coordinator.

S   ☐ Appoint   or   acquire   hazard   mitigation
              coordinator.

P (a/p) ☐ 132. Issue   public   information   announcements,   as
                 necessary.

P (p)   ☐ 133. Activate the Recovery Information Hotline.

P (a/p/o) ☐ 134. Determine   if   a   curfew   will be   necessary for
                 damaged areas.

P (a)   ☐ 135. Discuss   with   law   enforcement   and   judicial
                 officials   how curfew   violators will   be handled
                 (e.g., is there sufficient space in the jail or
                 stockade to house them or will violators be given
                 a fine and assigned to work cleanup crews).

P (a/o) ☐ 136. Discuss   with   law   enforcement   and   judicial
                 officials   of   court   trials   (both   civil   and
                 criminal) should be temporarily discontinued and
                 if so, for how long).

EM ASSIGNMENT: (a) = Administration (p) = Planning (o) = Operations
               (r) = Resource Management (s) = Staff
LEVEL OF PRIORITY: P = Primary S = Secondary T = Tertiary

A-19

P (a/p) ☐ 137. Determine method of assessing damages.

P (a/o) ☐ 138. Activate damage assessment teams.

P (o)   ☐ 139. Conduct and coordinate debris clearance.

T (a/o) ☐ 140. Acquire appropriate permits or permission for debris removal and disposal.

P (a/o) ☐ 141. Conduct and coordinate damage assessments.

P (a/s) ☐ 142. Establish portage areas.

P (a)   ☐ 143. Acquire funds to purchase needed emergency resources.

P (r)   ☐ 144. Monitor public health conditions and correct deficiencies.

P (a/s) ☐ 145. Evaluate the long-term commitment needed for capital facilities planning.

RESTORATION   * * * * * * * * 26 RESPONSE ACTIONS * * * * * * * *

P (a/s)   ☐ 146. Perform assessment of community needs.

P (a/r)   ☐ 147. Coordinate emergency relief assistance.

P (a/p/r) ☐ 148. Establish emergency worker stations and coordinate support activities.

P (p/r)   ☐ 149. Establish staging areas.

P (p/r)   ☐ 150. Establish recovery centers and coordinate support activities.

P (p/r)   ☐ 151. Establish emergency distribution centers and coordinate support activities.

S (a)   ☐ 152. Establish and maintain a designated Federal Public Assistance Office and coordinate activities.

P (a)   ☐ 153. Attend the public officials' briefing-Federal Public Assistance.

EM ASSIGNMENT: (a) = Administration (p) = Planning (o) = Operations
               (r) = Resource Management (s) = Staff
LEVEL OF PRIORITY: P = Primary   S = Secondary   T = Tertiary

P　(r)　☐　154. Provide all law enforcement agencies with price lists of needed items (e.g., generators, chain saws, ice, etc.) from reputable vendors so that these agencies can enforce county price gouging ordinance.

P　(a/s)　☐　155. Complete the Federal Public Assistance-Notice of Interest Form.

P　(a)　☐　156. Enact a Resolution designating the Applicant's Agent for Federal/State Assistance.

P　(a/o)　☐　157. Activate the appropriate members of the Damage Survey Team.

T　(p)　☐　158. Collect and complete appropriate reports and submit summary to State DEM.

S　(p)　☐　159. Collect and compile the following reports:

S　　　　☐　Daily Activity

S　　　　☐　Action/Event Logs

S　　　　☐　Data on damage eligible for Federal reimbursement.

P　(p/o)　☐　160. Provide assistance in the establishment and coordination of the Federal Damage Survey Reports.

T　(a/r)　☐　161. Provide assistance in the establishment, staffing and operations of Disaster Field Offices (DFOs).

S　(a/r)　☐　162. Provide assistance in the establishment, staffing and operations of Disaster Application Centers (DACs).

T　(a/r)　☐　163. Provide assistance in the establishment of temporary housing sites.

P　(a/p)　☐　164. Complete Federal Project Applications.

T　(a/s)　☐　165. Complete the following:

T　　　　☐　After Evacuation Report

T　　　　☐　County Incident Profile Report

EM ASSIGNMENT: (a) = Administration (p) = Planning (o) = Operations
　　　　　　　　(r) = Resource Management (s) = Staff
LEVEL OF PRIORITY:　P = Primary　S = Secondary　T = Tertiary

APPENDIX E

P (a/s) ☐ 166. Critique the management of the storm emergency.

P (a/s) ☐ 167. With assistance from State and Federal agencies, assess the County and its municipalities ' emergency management programs.

S (a/p) ☐ 168. Provide assistance in the establishment and coordination of State/Federal hazard mitigation efforts.

S (a/p) ☐ 169. Review and examine existing construction practices, future growth policies and development practices.

S (a/p) ☐ 170. Review and/or develop hazard mitigation policies and/or standards.

S (a/p) ☐ 171. Propose local laws to mitigate hurricane hazard damages.

RECONSTRUCTION * * * * * * * * * 5 RESPONSE ACTIONS * * * * * * * *

P (a/p) ☐ 172. Perform long-term activities or projects focused on improving or strengthening the community's economy.

P (a/p) ☐ 173. Perform hazard mitigation projects or programs to reduce the community's hurricane susceptibility and vulnerability.

P (a/p) ☐ 174. Repair, replace, modify or relocate public facilities in hazard-prone areas.

P (a/p) ☐ 175. Develop and implement a redevelopment plan for hazard-prone areas that would minimize repeated exposure to life-threatening situations.

P (a/p) ☐ 176. Implement a acquisition program to acquire storm-damage property in hazard-prone areas.

EM ASSIGNMENT: (a) = Administration (p) = Planning (o) = Operations
(r) = Resource Management (s) = Staff
LEVEL OF PRIORITY: P = Primary S = Secondary T = Tertiary

APPENDIX F

Units

Unfortunately, the units used to measure wind speed, temperature, pressure and so forth are not standard within the United States or between the United States and the rest of the world. Even scientists switch back and forth between units. Wind speed, for example, is expressed in meters per second, kilometers per hour, statute miles per hour, and nautical miles per hour (knots).

To assist the reader, we provide the conversion list below. In the text, we have sought to use English units with a metric equivalent in parentheses. While there are exceptions to this order in the text (and in the Figures), we have adopted English units as the primary set of units because most of the readers are expected to be most familiar with them. Also to standardize the text as much as possible, we have tried to avoid knots and use statute miles per hour (when we use "miles per hour" this means statute miles per hour).

The conversions are as follows.

WIND SPEED

1 nautical mile per hour (1 knot) = 1.15 statute miles per hour = 0.515 meters per second
1 statute mile per hour = 0.447 meters per second
1 meter per second = 2.233 miles per hour

DISTANCE

1 meter = 3.281 feet
1 foot = 0.3048 meters
1.609 kilometers = 1 mile
2.54 centimeters = 1 inch

TEMPERATURE

1°F = 5/9°C
1°C = 9/5°F
0°C = 32°F

A NOTE ON DOLLARS

When discussing dollars, we have identified when we are using constant versus current year dollars. When using constant year dollars, we have identified the base year.

References

Aberson, S.D. and M. DeMaria, 1994: Verification of a nested barotropic hurricane track forecast model (VICBAR). *Mon. Wea. Rev.*, **75**, 2804–2815.

Achter, S.M. and A.W. McGowan, 1984: *Land Use Planning for Natural Hazards: An Annotated Bibliography*, CPL Bibliographies, Chicago, IL, 32 pp.

Ager, D., 1993: *The New Catastrophism*, Cambridge University Press, Cambridge, UK, 231 pp.

Alexander, D., 1993: *Natural Disasters*, Chapman and Hall, New York.

Alvarez, L., 1992a: Storm at most savage in 2 South Dade cities. *The Miami Herald*, August 25, p. 1A.

Alvarez, L., 1992b: Job-seekers flood S. Dade, seek gold amid ruins. *The Miami Herald*, September 10, p. 1B.

AMS (American Meteorological Society), 1986: Is the United States headed for hurricane disaster? A statement of concern. *Bull. Am. Meteor. Soc.*, **67**, 537–538.

AMS (American Meteorological Society), 1993: Hurricane detection, tracking and forecasting. *Bull. Am. Meteor. Soc.*, **74**, 1377–1380.

Anderson, M.B., 1995: Vulnerability to disaster and sustainable development: A general framework for assessing vulnerability. In: *Disaster Prevention for Sustainable Development: Economic and Policy Issues*, M. Munasinghe and C. Clarke, Eds., IDNDR and The World Bank, Washington, DC.

Anderson, P., G. Lim and M. Merzer, 1992: Chiles seeks $9 billion in aid, Bush will urge $7.2 billion package. *The Miami Herald*, September 9, p. 1A.

Anthes, R., 1982. *Tropical Cyclones: Their Evolution, Structure and Effects*, Meteorological Monographs, Volume 19, American Meteorological Society, Boston, MA, 208 pp.

Anthes, R.A., and J.W. Trout, 1971: Three-dimensional particle trajectories in a model hurricane. *Weatherwise*, **24**, 174–178.

Applebome, P., 1993: In the Hurricane belt, a new, wary respect. *New York Times*, August 18, p. A12.

ATFHCI (Academic Task Force on Hurricane Catastrophe Insurance), 1995: Restoring Florida's paradise. State of Florida, 30 September 1995.

Avila, L.A. and R.J. Pasch, 1995: Atlantic tropical systems of 1993. *Mon. Wea. Rev.*, **123**, 887–896.

Baker, E.J. (Ed.), 1980: *Hurricanes and coastal storms*. Report Number 33. Florida Sea Grant College Program, University of Florida, Gainesville, FL.

Baker, E.J., 1993a: Emergency preparedness and public response in Southeast Florida in Andrew. In: *Lessons of Hurricane Andrew*, L.S. Tait, Ed., Excerpts from the 15th Annual National Hurricane Conference, 13–16 April, FEMA, Washington, DC.

Baker, E.J., 1993b: Empirical studies of public response to tornado and hurricane warnings in the United States. In: *Prediction and Perception of Natural Hazards*, J.

Nemec, J.M. Nigg and F. Siccardi, Eds., Kluwer Academic Publishers, Dordrecht, The Netherlands, 65–73.

Baker, E.J., 1994: Warning and response. In: *Hurricane Hugo: Puerto Rico, The Virgin Islands, and Charleston, South Carolina, September 17–22, 1989*, Natural Disaster Studies Volume 6, National Academy of Sciences, Washington, DC, 202–210.

Banham, R., 1993: Reinsurers seek relief in computer predictions. *Risk Mgt.*, August, 14–19.

Bardwell, L.V., 1991: Problem-framing: A perspective on environmental problem-solving. *Environ. Mgt.*, **15**, 603–612.

Barry, J., 1992: The shelter experience. *The Miami Herald*, August 25, p. 1B.

Baumann, D.D. and J.H. Sims, 1974: Human response to the hurricane. In: *Natural Hazards: Local, National, Global*, G. F. White, Ed., Oxford University Press, London, Chapter 3, 25–30.

Bender, M.A., R.E. Tuleya, and Y. Kurihara, 1987: A numerical study of the effect of island terrain on tropical cyclones. *Mon. Wea. Rev.*, **115**, 130–155.

Bender, M.A., R.J. Ross, R.E. Tuleya, and Y. Kurihara, 1993: Improvements in tropical cyclone track and intensity forecasts using the GFDL initialization system. *Mon. Wea. Rev.*, **121**, 2046–2061.

Benenson, B., 1993: Members seeking budget cuts bottle up flood relief. *Congressional Quarterly*, **51**, 1941–1943.

Benktander, G. and B. Berliner, 1977: Risk and return in insurance and reinsurance. *J. Risk Insurance*, **44**, 299–304.

Black, P.G. and R.M. Wakimoto, 1994: Damage survey of Hurricane Andrew and its relationship to the eyewall. *Bull. Am. Meteor. Soc.*, **75**, 189–200.

Blaikie, P.M., T. Cannon, I. Davis, and B. Wisner, 1994: *At Risk: Natural Hazards, People's Vulnerability and Disasters*, Routledge, London, 284 pp.

Bodeker, H., 1992: Channel 4, Norcross excel. *The Miami Herald*, August 25, p. 12 in "Living Today".

Bohle, H.G., T.E. Downing and M.J. Watts, 1994: Climate change and social vulnerability. *Global Environ. Change*, **4**, 37–48.

Boose, E.R., D.R. Foster, and M. Fluet, 1994: Hurricane impacts to tropical and temperate forest landscapes. *Ecol. Monogr.*, **64**, 369–400.

Bosart, L.F., 1984: The Texas coastal rainstorm of 17–21 September 1979: An example of synoptic-mesoscale interaction. *Mon. Wea. Rev.*, **112**, 1108–1133.

Brinkmann, W.A.R., 1975: Hurricane hazard in the United States: A research assessment. Monograph #NSF-RA-E-75-007, Program on Technology, Environment and Man, Institute of Behavioral Sciences, University of Colorado, Boulder, CO.

Brown, G., 1992: The objectivity crisis. *Am. J. Phys.*, **60**, 779–781.

Brunner, R.D., 1991: Global climate change: Defining the policy problems. *Policy Sci.*, **24**, 291–311.

Brunner, R.D. and W. Ascher, 1992: Science and social accountability. *Policy Sci.*, **25**, 295–331.

Bryant, E.A., 1991: *Natural Hazards*, Cambridge University Press, Cambridge.

BTFFDR (Bipartisan Task Force on Funding Disaster Relief), 1995: Federal Disaster Assistance: Report of the Senate Task Force on Funding Disaster Relief, 104–4. GPO, Washington, DC.

Burby, R.J. and L.C. Dalton, 1994: Plans can matter! The role of land use plans and state planning mandates in limiting the development of hazardous areas. *Public Admin. Rev.*, **54**, 229–237.

Burpee, R.W., S.D. Aberson, P.G. Black, M. DeMaria, J.L. Franklin, J.S. Griffin, S.H. Houston, J. Kaplan, S.J. Lord, F.D. Marks, Jr, M.D. Powell and H.E.

260   HURRICANES: THEIR NATURE AND IMPACT ON SOCIETY

Willoughby, 1994: Real-time guidance provided by NOAA's Hurricane Research Division to forecasters during Emily of 1993. *Bull. Am. Meteor. Soc.*, **75**, 1765–1783.

Burpee, R.W., J.L. Franklin, S.J. Lord, R.E. Tuleya and S.D. Aberson, 1996: The impact of Omega dropwindsondes on operational hurricane track forecast models. *Bull. Am. Meteor. Soc.*, **77**, 925–934.

Burton, I., R.W. Kates and G.F. White, 1993: *The Environment as Hazard*, 2nd edition, Guilford Press, New York.

Byerly, R., 1995: U.S. science in a changing context: A perspective. In *US National Report to the International Union of Geodesy and Geophysics 1991–1994*, R.A. Pielke, Ed., American Geophysical Union, Washington, DC, A1-A16.

Byerly, R., Jr and R.A. Pielke, Jr, 1995: The changing ecology of United States science. *Science*, **269**, 1531–1532.

Byers, H.R., 1974: History of weather modification. In: *Weather and Climate Modification*, W.N. Hess, Ed., John Wiley and Sons, New York, 3–44.

Camerer, C.F. and H. Kunreuther, 1989: Decision process for low probability events: Policy implications. *J. Policy Anal. Manage.*, **8**, 565–592.

Campbell, P.R., 1994: Population projections for states, by age, sex, race, and Hispanic origin: 1993 to 2020. Current Population Reports, P25-1111. US GPO, Washington, DC.

Carter, T.M., 1993: The role of technical hazard and forecast information in preparedness for and response to the hurricane hazard in the United States. In: J. Nemec, J.M. Nigg and F. Siccardi, Eds., *Prediction and Perception of Natural Hazards*, Kluwer Academic Publishers, Dordrecht, The Netherlands, 75–81.

Chambers, R., 1989: Vulnerability, coping and policy. *IDS Bull.*, **20**, 1–7.

Chan, J.C.L. and W.M. Gray, 1982: Tropical cyclone movement and surrounding flow relationships. *Mon. Wea. Rev.*, **110**, 1354–1374.

Chan, J.C.L. and J.-E. Shi, 1996: Long-term trends and interannual variability in tropical cyclone activity over the western North Pacific. *J. Geophys. Res. Lett.*, **23**, 2765–2767.

Chang, S., 1984: Do disaster areas benefit from disasters? *Growth Change*, **15**(4), 24–31.

Changnon, S.A., 1975: The paradox of planned weather modification. *Bull. Am. Meteor. Soc.*, **56**, 27–37.

Changnon, S., 1994: Impacts: Overview. In: N.G. Bhowmik et al., Eds., *The 1993 Flood on the Mississippi River in Illinois*, Illinois Water Survey, Miscellaneous Publication 151. Champaign, IL, 86–87.

Changnon, S.A. and J.M. Changnon, 1992: Temporal fluctuations in weather disasters: 1950–1989. *Climatic Change*, **22**, 191–208.

Changnon, S.A., J.M. Changnon and D. Changnon, 1995: Uses and applications of climate forecasts for power utilities. *Bull. Am. Meteor. Soc.*, **76**, 711–720.

Changnon, S.A., D. Changnon, E.R. Fosse, D.C. Hoganson, R.J. Roth, Sr, and T. Totsch, 1996: Impacts and responses of the weather insurance industry to recent weather extremes. Final Report to the University Corporation for Atmospheric Research, CRR-41, May 1996.

Clark, W.C. and G. Majone, 1985: The critical appraisal of scientific inquiries with political implications. *Sci. Technol., Hum. Values*, **10**(3), 6–19.

Cobb, H.D., III, 1991: The Chesapeake-Potomac hurricane of 1933. *Weatherwise*, **44**, 24–29.

Coch, N.K., 1994: Geologic effects of hurricanes. *Geomorphology*, **10**, 37–63.

Coch, N.K. and M.P. Wolff, 1990: Probable effects of a storm like Hurricane Hugo on Long Island, New York. *Northeastern Environ. Sci.*, **9**, 33–47.

Cochran, L. and M. Levitan, 1994: Lessons from Hurricane Andrew. *Architect Sci. Rev.*, **37**, 115–121.

Corddry, M., 1991: *City on the Sand: Ocean City, Maryland, and the People who Built it*, Tidewater Publishers, Centreville, MD.

Cotton, W.R. and R.A. Pielke, 1995: *Human Impacts on Weather and Climate*, Cambridge University Press, New York, 288 pp.

Council of Economic Advisors, 1994: Economic Report of the President. US GPO, Washington, DC, (February).

Davis, W.P., K.W. Thornton and B. Levinson, 1994: Framework for assessing effects of global climate change on mangrove ecosystems. *Bull. Marine Sci.*, **54**, 1045–1058.

DeAngelis, R.M. and W.T. Hodge, 1972: Preliminary climatic data report Hurricane Agnes June 14–23, 1972. NOAA Technical Memorandum EDS NCC-1, August, National Climatic Center, Asheville, NC.

DeMaria, M., 1995: A history of hurricane forecasting for the Atlantic Basin, 1920–1995. In: *Meteorology since 1919: Essays Commemorating the 75th Anniversary of the American Meteorological Society*. American Meteorological Society, Boston, MA.

DeMaria, M., 1996: The effect of vertical shear on tropical cyclone intensity change. *J. Atmos. Sci.*, **53**, 2076–2087.

DeMaria, M. and J. Kaplan, 1994: A statistical hurricane intensity prediction scheme (SHIPS) for the Atlantic Basin. *Wea. Forecasting*, **9**, 209–220.

DeMaria, M., M.B. Lawrence and J.T. Kroll, 1990: An error analysis of Atlantic tropical cyclone track guidance models. *Wea. Forecasting*, **5**(3), 47–61.

Dewar, H., 1992: Devastation hits home in S. Fla. *The Miami Herald*, August 24, Special Hurricane Edition, p. 4.

Dewey, J., 1933: John Dewey: The later works, 1925–1953. Southern Illinois University Press, Carbondale and Edwardsville, 113–139.

Diamond, H.L. and P.F. Noonan, Eds. 1996: *Land use in America*, Island Press, Washington, DC, 351 pp.

Diaz, H. and R. Pulwarty, 1997: *Hurricanes: Climate, Social, and Economic Impacts*, Springer-Verlag, New York.

Dlugolecki, A.F., K.M. Clark, D. McCauley, J.P. Palutikof and W. Yambi, 1994: IPCC – Second Assessment Report, Working Group II: Impacts of Climate Change, Chapter B9, Financial Services (First Draft). IPCC, Geneva, Switzerland.

Dlugolecki, A.F., K.M. Clark, F. Knecht, D. MacCauley, J.P. Palutikof and W. Yambi, 1996: Financial services. In: *Climate Change 1995: Impacts, Adaptations and Mitigation of Climate Change: Scientific–Technical Analyses*. Contribution of Working Group II to the Second Assessment Report of the Intergovernmental Panel on Climate Change. Cambridge University Press, London, chapter 17.

DOC (Department of Commerce) 1964: *Hurricane Dora, August 28-September 16, 1964*, Preliminary Report with Advisories and Bulletins Issued, Washington, DC.

DOC (Department of Commerce), 1990: Hurricane Hugo September 10–22, 1989. Natural Disaster Survey Report. National Weather Service, Silver Spring, MD (May).

DOC (Department of Commerce), 1993: Natural Disaster Survey Report, Hurricane Andrew: South Florida and Louisiana, August 23–26, 1992. National Weather Service, Silver Spring, MD, (November).

Doehring, F., I.W. Duedall and J.M. Williams, 1994: Florida hurricanes and tropical storms. 1871–1993: An historical survey. Florida Sea Grant College Program TP-71. University of Florida, Gainesville, FL.

Doig, S.K., 1992a: The ocean's fury. *The Miami Herald*, October 1, p. 1D.

262   HURRICANES: THEIR NATURE AND IMPACT ON SOCIETY

Doig, S.K., 1992b: Storm: A wobble away from great disaster. *The Miami Herald*, September 6, p. 23A.

Donnelly, J., 1992: Andrew left our defenses down. *The Miami Herald*, September 4, p. 22A.

Downing. T.E., 1991: Vulnerability to hunger in Africa. *Global Environ. Change*, 1, 365–380.

Downing, T.E., 1992: Climate change and vulnerable places: Global food security and country studies in Zimbabwe, Kenya, Senegal and Chile. Research Report 1. Environmental Change Unit, University of Oxford, UK.

Downs, A., 1972: Up and down with ecology – the "issue-attention cycle." *The Public Interest*, **28**, 38–50.

Doyle, T.W. and G.F. Girod, 1997: The frequency and intensity of Atlantic hurricanes and their influence on the structure of south Florida mangrove communities, pp. 109–120. In H.D. Diaz and R Pulwarty Eds, *Hurricanes: Climate and Socioeconomic Impacts*, Springer, Heidelburg, Germany.

Drabek, T., 1986: *Human System Responses to Disaster*. Springer-Verlag, New York.

Dubocq, T., 1992: Power FPL uncertain about when all service will be restored. *The Miami Herald*, August 26, p. 1A.

Dunn, G.E. and B.I. Miller, 1964: *Atlantic Hurricanes*, Louisiana State University Press, Baton Rouge, LA, 377 pp.

Dvorak, V.F., 1975: Tropical cyclone intensity analysis and forecasting from satellite imagery. *Mon. Wea. Rev.*, **103**, 420–430.

Dvorak, V.F., 1984: Tropical cyclone intensity analysis using satellite data. NOAA Technical Report NESDIS 11, 47 pp.

Eagleman, J.R., 1990: *Severe and Unusual Weather*, 2nd edition, Trimedia Publishing, Lenexa, KS.

Eastman, J.L., 1995: Numerical simulation of Hurricane Andrew – Rapid intensification. *21st Conference on Hurricanes and Tropical Meteorology*, AMS, Boston, 24–28 April, Miami, FL, 111–113.

Elgiston, H., 1992: Homestead: Ground zero. In J. Macchio Ed., *Hurricane Andrew: Path of Destruction*, August 1992, BD Publishing, Charleston, SC.

Elsberry, R.L., 1995: Recent advancements in dynamical tropical cyclone track predictions. *Meteor. Atmos. Phys.*, **56**, 81–99.

Elsberry, R.L. and R.A. Jeffries, 1996: Vertical wind shear influences on tropical cyclone formation and intensification during TCM-92 and TCM-93. *Mon. Wea. Rev.*, **124**, 1374–1387.

Elsberry, R.L., W.M. Frank, G.J. Holland, J.D. Jarrell and R.L. Southern, 1987: A global view of tropical cyclones. Based largely on materials prepared for the International Workshop on Tropical Cyclones, Bangkok, Thailand, 25 November–5 December 1985. Office of Naval Research, Marine Meteorology Program, 185 pp.

Emanuel, K.A., 1986: An air–sea interaction theory for tropical cyclones. Part I. *J. Atmos. Sci.*, **42**, 586–604.

Emanuel, K.A., 1987: The dependence of hurricane intensity on climate. *Nature*, **326**, 483–485.

Emanuel, K.A., 1988: Toward a general theory of hurricanes. *Am. Sci.*, **76**, 371–379.

Emanuel, K.A., 1996: Maximum hurricane intensity estimation. Available at http://cirrus.mit.edu/~emanuel/pcmin/pclat/pclat.html

Emanuel, K.A., K. Speer, R. Rotunno, R. Srivastava and M. Molina, 1995: Hypercanes: a possible link in global extinction scenarios. *J. Geophys. Res.*, **100**, 13 755–13 765.

Englehardt, J. and C. Peng, 1996: A Bayesian benefit–risk model applied to the south Florida building code. *Risk Anal.*, **16**(1), 81–91.

Etzioni, A., 1985: Making policy for complex systems: A medical model for economics. *J. Policy Anal. Manage.*, **4**, 383–395.

Fatsis, S., 1992: Andrew reshapes S. Florida economy. *The Miami Herald*, September 8, Special Section, p. 10.

FDCA (Florida Department of Community Affairs), 1992: Hurricane Andrew: Recommendations for building codes and building code enforcement. Final Report prepared by Ronald A. Cook, University of Florida, DCA Contract No. 93RD-66-13-00-22-001.

Feldman, M.S. and J.G. March, 1981: Information in organizations as signal and sign. *Admin. Sci. Quart.*, **26**, 171–186.

FEMA (Federal Emergency Management Agency), 1993: FEMA's Disaster Management Program: A performance audit after Hurricane Andrew, H-01-93, January. FEMA, Washington, DC.

FEMA/FIA (Federal Emergency Management Agency/Federal Insurance Administration), 1992: Building performance: Hurricane Andrew in Florida, FIA-22, December 21. FEMA, Washington, DC.

FEMA/FIA (Federal Emergency Management Agency/Federal Insurance Administration), 1993: Building performance: Hurricane Andrew in Florida. FEMA, Washington, DC.

Fielder, T., 1992: Messages of hope storm drives home the need for leadership. *The Miami Herald*, August 30, p. 1L.

Fields, G., 1992: Andrew pulls dollars from wallets. *The Miami Herald*, September 26, p. 1C.

Fields, G., 1994: Cover story: Florida's economy '94 outlook. *The Miami Herald*, January 17, p. 17BM.

FIFMTF (Federal Interagency Floodplain Management Task Force), 1992: Floodplain management in the United States: An assessment report, Volume 2: Full Report, L.R. Johnston Associates.

Filkins, D., 1992: Planners forecast 2 futures for Dade. *The Miami Herald*, October 19, p. 1A.

Filkins, D., 1994: Plan would raise most Dade tax bills, budget also proposes increased water fees. *The Miami Herald*, July 16, p. 1A.

Finefrock, D., 1992: Legislature could ease property tax burden. *The Miami Herald*, September 6, p. 10G.

Fischer, A., L.G. Chestnut and D.M. Violette, 1989: The value of reducing risks of death: A note on new evidence. *J. Policy Anal. Manage.*, **8**, 88–100.

Fischhoff, B., S.R. Watson and C. Hope, 1984: Defining risk. *Policy Sci.*, **17**, 123–139.

FPMA (Floodplain Management Assessment), 1995: A floodplain management assessment of the upper Mississippi River and lower Missouri rivers and tributaries. U.S. Army Corps of Engineers, St Louis, Missouri.

Forester, J., 1984: Bounded rationality and the politics of muddling through. *Public Admin. Rev.*, **44**, 23–31.

Forsythe, J.M. and T.H. Vonder Haar, 1996: A warm core in a polar low observed with a satellite microwave sounding unit. *Tellus*, **48A**, 193–208.

Foster, D.R., 1988: Species and stand response to catastrophic wind in central New England, U.S.A. *J. Ecol.*, **76**, 135–151.

Foster, D.R., and E.R. Boose, 1992: Patterns of forest damage resulting from catastrophic wind in central New England. *USA J. Ecol.*, **80**, 79–98.

Foster, D.R. and E.R. Boose, 1995: Hurricane disturbance regimes in temperate and tropical forest ecosystems. In: *Wind and Trees*, M.P. Coutts and J. Grace, Eds, Cambridge University Press, London, 305–339.

Fujita, T., 1992: Damage survey of Hurricane Andrew in south Florida. *Storm Data*, **34**(8), 24–29.

GAO (General Accounting Office), 1993a: Disaster relief fund: Actions still needed to prevent recurrence of funding shortfall. GAO/RCED-93-60. General Accounting Office, Washington, DC.

GAO (General Accounting Office), 1993b: Disaster management: Improving the nation's response to catastrophic disasters. GAO/RCED-93-186. General Accounting Office, Washington, DC.

GAO (General Accounting Office), 1993c: Disaster Assistance: DOD's Support for Hurricanes Andrew and Iniki and Typhoon Omar. GAO/NSIAD-93-180 (June), 10–11. General Accounting Office, Washington, DC.

Garcia, B. and D. Satterfield, 1992: Insurer called storm "opportunity" to raise its rates. *The Miami Herald*, September 5, 1A.

Garcia, M., D. Neal and J. Tanfani, 1992: Andrew's death list spanned Dade County. *The Miami Herald*, August 26, 23A.

Gentry, R.C., 1974: Hurricane modification. In: *Weather and Climate Modification*, W.N. Hess, Ed., John Wiley and Sons, New York, 497–521.

George, P.S., 1992: Bryan Norcross: A hero in the wake of Hurricane Andrew. In: *Hurricane Andrew: Path of Destruction*, J. Macchio, Ed., BD Publishing, Charleston, SC.

Getter, L., 1992a: Why don't people evacuate? Denial, turf worries. *The Miami Herald*, August 25, 12.

Getter, L., 1992b: Inspections: a breakdown in the system. *The Miami Herald*, December 20, 65R.

Getter, L., 1992c: Do builders' bucks buy political power? *The Miami Herald*, December 20, 75R.

Getter, L., 1993: Inspections: A breakdown in the system. In: *Lessons of Hurricane Andrew*, Tait, L.S. Ed., Excerpts from the 15th Annual National Hurricane Conference. FEMA (April), Washington, DC, 33–36.

Gibbs, W.J., 1987: Defining climate. *WMO Bull.*, **36**, 290–296.

Glantz, M.H., 1977: The value of a long-range weather forecast for the West African Sahel. *Bull. Am. Meteor. Soc.*, **38**, 150–158.

Glantz, M.H., 1978: Render unto weather . . . – an editorial. *Climatic Change*, **1**, 305–306.

Glantz, M.H., 1979: *Saskatchewan Spring Wheat Production 1974: A Preliminary Assessment of a Reliable Long-Range Forecast*, Climatological Studies No. 33, Canadian Government Publishing Centre, Hull, Quebec, Chapter 8.

Glantz, M.H., 1982: Consequences and responsibilities in drought forecasting: The case of Yakima, 1977. *Water Resources Res.*, **18**, 3–13.

Glantz, M.H., 1986: Politics, forecasts, and forecasting: Forecasts are the answer, but what was the question? In: *Policy Aspects of Climate Forecasting*, R. Krasnow, Ed., Resources for the Future, Washington, DC.

Glantz, M.H., 1988: Societal Responses to Regional Climatic Change: Forecasting by Analogy. Westview Press, Boulder, CO.

Glantz, M.H., 1996: *Currents of Change: El Niño's Impact on Climate and Society*. Cambridge University Press, Cambridge, UK.

Golden, J.H., 1990: Meteorological data from Hurricane Hugo. *Proc. 22nd Joint UJNR Panel Meetings – Winds and Seismic Effects* (May 14–18). National Institute of Standards and Technology, Gaithersburg, MD.

Goldenberg, S.B. and L.J. Shapiro, 1996: Physical mechanisms for the association of El Niño and West African rainfall with Atlantic major hurricane activity. *J. Climate*, **9**, 1169–1187.

Graham, B., 1992: Andrew's lesson, a redefined role for the military. *The Miami Herald*, September 6, 4M.

Gray, W.M., 1968: A global view of the origin of tropical disturbance and storms. *Mon. Wea. Rev.*, **96**, 669–700.

Gray, W.M., 1990: Strong association between West African rainfall and U.S. landfall of intense hurricanes. *Science*, **249**, 1251–1256.

Gray, W.M., 1992: Summary of 1992 Atlantic tropical cyclone activity and verification of author's forecast. Report of 24 November 1992, Department of Atmospheric Science, Colorado State University, Fort Collins, CO.

Gray, W.M., 1994: Extended range forecast of Atlantic seasonal hurricane activity for 1995. Department of Atmospheric Science, Colorado State University, Fort Collins, CO.

Gray, W.M., 1995: Early April assessment of the forecast of Atlantic Basin seasonal hurricane activity for 1995. Department of Atmospheric Science, Colorado State University, Fort Collins, CO.

Gray, W.M., 1997: Tropical cyclones. Report prepared for the World Meteorological Organization, Geneva, Switzerland, 164 pp., in press.

Gray, W.M. and C.W. Landsea, 1992: African rainfall as a precursor of hurricane-related destruction on the U.S. East Coast. *Bull. Am. Meteor. Soc.*, **73**, 1352–1364.

Gray, W.M., C. Neumann and T.L. Tsui, 1991: Assessment of the role of aircraft reconnaissance on tropical cyclone analysis and forecasting. *Bull. Am. Meteor. Soc.*, **72**, 1867–1883.

Gray, W.M., C.W. Landsea, P.W. Mielke, Jr and K.J. Berry, 1995: Early August updated forecast of Atlantic seasonal hurricane activity for 1995. Colorado State University, Fort Collins, CO, 26 pp.

Haas, J.E., R.W. Kates, and M.J. Bowden, Eds., 1977: *Reconstruction Following Disaster*. MIT Press, Cambridge, MA.

Hancock, D. and A. Faiola, 1992: Bigger, stronger, closer. *The Miami Herald*, August 23, 1A.

Haner, J. and K. Rafinski, 1992: Broward's emergency plan praised despite glitches. *The Miami Herald*, August 26, 1BR.

Hartman, T., 1992: S. Dade's last tent city folds up. *The Miami Herald*, October 24, 1A.

Hebert, P.J., J.D. Jarrell and M. Mayfield, 1993: The deadliest, costliest, and most intense United States hurricanes of this century (and other frequently requested hurricane facts). NOAA Technical Memorandum NWS NHC-31, February 1993, National Hurricane Center, Coral Gables, FL.

Hebert, P.J., J.D. Jarrell, and M. Mayfield, 1996: The deadliest, costliest, and most intense United States hurricanes of this century (and other frequently requested hurricane facts). NOAA Technical Memorandum NWS TPC-1, February 1996, National Hurricane Center, Miami, FL.

Henry, J.A., K.M. Portier and J. Coyne, 1994: *The Climate and Weather of Florida*, Pineapple Press, Sarasota, FL.

Hess, J.C. and J.B. Elsner, 1994: Historical developments leading to current forecast models of annual Atlantic hurricane activity. *Bull. Am. Meteor. Soc.*, **75**, 1611–1621.

Hewitt, K., 1983: *Interpretations of Calamity from the Viewpoint of Human Ecology*, Allen & Unwin, Boston, MA.

Higham, S., 1992a: Dash for FEMA cash yields dubious claims. *The Miami Herald*, October 26, p. 1A.

Higham, S., 1992b: Last of U.S. military pulls out of storm area. *The Miami Herald*, October 16, p. 1B.

Hilgartner, S. and C. Bosk, 1988: The rise and fall of social problems: A public arenas model. *Am. J. Sociol.*, **94**, 53–78.

Ho, F.P., J.C. Su, K.L. Hanevich, R.J. Smith and F.P. Richards, 1987: Hurricane climatology for the Atlantic and Gulf Coasts of the United States. NOAA Technical Memorandum NWS-38 (April). NWS, Silver Springs, MD.

Holland, G.J., 1983: Tropical cyclone motion: Environmental interaction plus a beta effect. *J. Atmos. Sci.*, **40**, 328–342.

Holland, G.J., 1993a: Ready reckoner. In: *Global Guide to Tropical Cyclone Forecasting*. World Meteorological Organization Technical Document, WMO/TD No. 560, Tropical Cyclone Programme, Report No. TCP-31, Geneva, Switzerland, Chapter 9.

Holland, G.J., 1993b: Tropical cyclone motion. In: *Global Guide to Tropical Cyclone Forecasting*. World Meteorological Organization Technical Document, WMO/TD No. 560, Tropical Cyclone Programme, Report No. TCP-31, Geneva, Switzerland, Chapter 3.

Holland, G.J., 1997: The maximum potential intensity of tropical cyclones. *J. Atmos. Sci.*, in press.

HRD (Hurricane Research Division), 1994: Fiscal year 1994 programs – Fiscal year 1995 plans. NOAA Atlantic Oceanographic and Meteorological Laboratory, Miami, FL.

IIPLR (IRC) (Insurance Institute for Property Loss Reduction and Insurance Resources Council), 1995: *Coastal Exposure and Community Protection: Hurricane Andrew's Legacy*, Boston, MA.

IPCC (Intergovernmental Panel on Climate Change), 1996a: *Climate Change 1995: The Science of Climate Change*. Contribution of Working Group I to the Second Assessment Report of the IPCC, Cambridge University Press, London, England.

IPCC (Intergovernmental Panel on Climate Change), 1996b: *Climate Change 1995: The Science of Climate Change*. Contribution of Working Group I to the Second Assessment Report of the IPCC, Cambridge University Press, London.

IPCC (Intergovernmental Panel on Climate Change), 1996c: *Climate Change 1995: Impacts, Adaptations, and Mitigation of Climate Change, Scientific–Technical Analyses*. Contribution of Working Group II to the Second Assessment Report of the IPCC, Cambridge University Press, London.

Jarrell, J.D., P.J. Hebert and M. Mayfield, 1992: Hurricane experience levels of coastal county populations from Texas to Maine. NOAA Technical Memorandum NWS NHC-46 (August). NHC, Coral Gables, FL.

Jarvinen, B.R. and M.B. Lawrence, 1985: An evaluation of the SLOSH storm-surge model. *Bull. Am. Meteor. Soc.*, **66**, 1408–1411.

Jelesnianski, C.P., 1993: The habitation layer. In: *Global Guide to Tropical Cyclone Forecasting*, G. Holland, Ed., WMO/TD – NO. 560, Report No. TCP-31, World Meteorological Society, Geneva.

Joe, P., C. Crozier, N. Donaldson, D. Etkin, E. Brun, S. Clodman, J. Abraham, S. Siok, H.-P. Biron, M. Leduc, P. Chadwick, S. Knott, J. Archibald, G. Vickers, S. Blackwell, R. Drouillard, A. Whitman, H. Brooks, N. Kouwen, R. Verret, G. Fournier and B. Kochtubajda, 1995: Recent progress in the operational forecasting of summer severe weather. *Atmos.-Ocean*, **33**, 249–302.

Kaplan, T.J., 1986: The narrative structure of policy analysis. *J. Policy Anal. Manage.*, **5**, 761–778.

Kaplan, J. and M. DeMaria, 1995: A simple empirical model for predicting the decay of tropical cyclone winds after landfall. *J. Appl. Meteor.*, **34**, 2499–2512.

Kates, R.W., 1980. Climate and society: Lessons from recent events. *Weather*, **35**, 17–25.

Katz, R.W., 1993: Towards a statistical paradigm for climate change. *Climate Res.*, **2**, 167–175.

Katz, R.W. and B.G. Brown, 1992: Extreme events in a changing climate: Variability is more important than averages. *Climatic Change*, **21**, 289–302.

Katz, R.W., 1993: Towards a statistical paradigm for climate change. *Climate Res.*, **2**, 167–175.

Kingdon, J., 1984: *Agendas, Alternatives, and Public Policies*, Harper Collins Publishers, New York.

Kleppner, D., 1993: Thoughts on being bad. *Physics Today*, **48**(10), 9–10.

Kocin, P.J. and J.H. Keller, 1991: A 100-year climatology of tropical cyclones for the northeast United States. Paper prepared for the 19th Conference on Hurricanes and Tropical Meteorology. American Meteorological Society, Miami, FL, 6–10 May.

Kotsch, W.J., 1977: *Weather for the Mariner*, 2nd ed., Naval Institute Press, Annapolis, MD.

Koutnik, F.J., 1993: Testimony before the Senate Environment and Public Works Committee. Hearing on Lessons Learned from Hurricane Andrew, S.Hrg. 103–86, April 1, 79–83. GPO, Washington, DC.

Landsea, C.W., 1991: West African monsoonal rainfall and intense hurricane association. Department of Atmospheric Science, Paper No. 484 (7 October), Colorado State University, Fort Collins, CO.

Landsea, C.W., 1993: A climatology of intense (or major) Atlantic hurricanes. *Mon. Wea. Rev.*, **121**, 1703–1713.

Landsea, C.W., 1996: Frequently asked questions report. Available at http://tropical.atmos.colostate.edu.

Landsea, C.W., W.M. Gray, P.W. Mielke, Jr and K.J. Berry, 1994: Seasonal forecasting of Atlantic hurricane activity. *Weather*, **49**, 273–284.

Landsea, C.W., N. Nicholls, W.M. Gray and L.A. Avila, 1996: Quiet early 1990s continues trend of fewer intense Atlantic hurricanes. *Geophys. Res. Lett.*, **23**, 1697–1700.

Lane, N., 1995: Statement before the House Committee on Science, Hearing on "Is Today's Science Policy Preparing Us for the Future?" January 6.

Lasswell, H.D., 1971: *A Pre-View of Policy Sciences*. American Elsevier, New York.

Lawrence, B.M. and J.M. Gross, 1989: Atlantic hurricane season of 1988. *Mon. Wea. Rev.*, **117**, 2248–2259.

Lawrence, J.R. and S.D. Gedzelman, 1996: Low stable isotope ratios of tropical cyclone rains. *Geophys. Res. Lett.*, **23**, 527–530.

Lawrence, M.B. and J.M. Pelissier, 1981: Atlantic hurricane season of 1980. *Mon. Wea. Rev.*, **109**, 1567–1582.

Leen, J., S.K. Doig, L. Getter, L.F. Soto and D. Finefrock, 1992: Failure of design and discipline. In: *Lessons of Hurricane Andrew*, L.S. Tait, Ed., Excerpts from the 15th Annual National Hurricane Conference. FEMA (April), Washington, DC, 39–44.

Leggett, J., Ed., 1994: *The Climate Time Bomb*, Greenpeace International, Amsterdam.

Leslie, L.M. and G.J. Holland, 1995: On the bogussing of tropical cyclones in numerical models: A comparison of vortex profiles. *Meteor. Atmos. Phys.*, **56**, 101–110.

Levy, L.J. and L.M. Toulman, 1993: Improving disaster planning and response efforts: Lessons from Hurricanes Andrew and Iniki. Prepared by Booz Allen & Hamilton, Inc., McLean, VA.

Lippmann, W., 1961: *Public Opinion*, Macmillan, New York, 427 pp.

Liu, G., J.A. Curry and C.A. Clayson, 1995: Study of tropical cyclogenesis using satellite data. *Meteor. Atmos. Phys.*, **56**, 111–123.

Liu, Y., D.L. Zhang and M.K. Yau, 1997: A multiscale numerical study of Hurricane Andrew (1992), Part I: Explicit simulation and verification. *Mon. Wea. Rev.* (in press).

Lodge, T.E., 1994: *The Everglades Handbook: Understanding the Ecosystem*, St Lucie Press, St Lucie, FL.

Lord, S.J., 1993: Recent developments in tropical cyclone track forecasting with the NMC global analysis and forecast system. *Preprints, 20th Conference on Hurricanes and Tropical Meteorology*, AMS, San Antonio, TX, 290–291.

Ludlam, D.M., 1963: *Early American Hurricanes: 1492–1870*, American Meteorological Society, Boston, MA, 198 pp.

Lyons, D., 1992: S. Dade curfew to end Monday. *The Miami Herald*, November 11, 1A.

Lyons, D. and M. Merzer, 1992: Answering an urgent cry: soldiers shocked by panorama of ruin. *The Miami Herald*, August 29, 1A.

Malkus, J.S., 1958: On the structure and maintenance of the mature hurricane eye. *J. Meteor.*, **15**, 337–349.

Malkus, J.S. and H. Riehl, 1960: On the dynamics and energy transformations in steady-state hurricanes. *Tellus*, **12**, 1–20.

Markowitz, A., 1992: An awful howl Andrew hits hardest in South Dade. *The Miami Herald*, August 24, Special hurricane Edition, 1.

Marks, D.G., 1992: The beta and advection model for hurricane track forecasting. NOAA Technical Memo. NWS NMC 70, National Meteorological Center, Camp Springs, MD, 89 pp.

Mauro, J., 1992: Hurricane Andrew's other legacy. *Psychol. Today*, **25**, 42–45, 93.

May, P., 1985: *Recovering from Catastrophes: Federal Disaster Relief Policy and Politics*, Greenwood Press, Westport, CT.

McAdie, C.J., 1991: A comparison of tropical cyclone track forecasts produced by NHC90 and an alternate version (NHC90A) during the 1990 hurricane season. *Preprints of the 19th Conference on Hurricanes and Tropical Meteorology*, Miami, AMS, Boston, MA, 290–294.

McAdie, C.J. and M.B. Lawrence, 1993: Long-term trends in National Hurricane Center Track Forecast errors in the Atlantic Basin. *Preprints, 20th Conference on Hurricanes and Tropical Meteorology*, AMS, San Antonio, TX, 281–284.

McAdie, C.J. and E.N. Rappaport, 1991: Diagnostic report of the National Hurricane Center, **4**, No. 1, NOAA, National Hurricane Center, Coral Gables, FL, 45 pp.

McElroy, C., 1996: On the causes of tropical cyclone motion and propagation. Department of Atmosperhic Science, MS Thesis, Colorado State University, Fort Collins, CO, 80523, 124 pp.

McNair, J., 1992a: Andrew's awesome impact. *The Miami Herald*, September 10, 4B.

McNair, J., 1992b: Study details hurricane's toll on local business. *The Miami Herald*, September 18, 1C.

McNew, K.P., H.P. Mapp, C.E. Duchon and E.S. Merritt, 1991: Sources and uses of weather information for agricultural decision makers. *Bull. Am. Meteor. Soc.*, **72**, 491–498.

Mehta, K.C., R.H. Cheshir and J.R. McDonald, 1992: Wind resistance categorization of buildings for insurance. *J. Wind Eng. Indust. Aerodyn.*, **43**, 2617–2628.

Merrill, R.T., 1985: Environmental influences on hurricane intensification. Department of Atmospheric Science, Paper No. 394, Colorado State University, Fort Collins, CO.

Merrill, R.T., 1987: An experiment in statistical prediction of tropical cyclone intensity change. NOAA Technical Memorandum NWS NHC-34, National Hurricane Center, Coral Gables, FL.

Mesthene, E.G., 1967: The impacts of science on public policy. *Public Admin. Rev.*, **27**, 97–105.

Miami Herald Staff, 1992a: What the grand juries said. *The Miami Herald*, September 13, 26A.

Miami Herald Staff, 1992b: Patchwork of destruction. *The Miami Herald*, December 20.

Miller, B.I., 1958: On the maximum intensity of hurricanes. *J. Meteor.*, **15**, 184–195.

Montgomery, M.T. and B.F. Farrell, 1993: Tropical cyclone formation. *J. Atmos. Sci.*, **50**, 285–310.

Morgenthaler, E., 1994: He's no blowhard: Dr. Gray can predict Atlantic hurricanes. *The Wall Street Journal*, August 8, A1.

Mrazik, B.R. and H. Appel Kinberg, 1991: National flood insurance program – Twenty years of progress toward decreasing nationwide flood losses. In: *National Water Summary 1988–89 – Hydrological Events and Floods and Droughts*, USGS Water Supply Paper 2375, 133–142.

Mulady, J.J., 1994: Building codes: They're not just hot air. *Natural Hazards Observer*, **18**(3), 4–5.

Murphy, A.H., 1993: What is a good forecast? An essay on the nature of goodness in weather forecasting. *Wea. Forecasting*, **8**, 281–293.

Murphy, A.H., 1994: Assessing the economic value of weather forecasts: An overview of methods, results, and issues. *Meteorological Appl.*, **1**, 69–73.

Namias, J., 1980: The art and science of long-range forecasting. *Eos*, **61**, 449–450.

NAS (National Academy of Sciences), 1980: Technological and scientific opportunities for improved weather and hydrological services in the coming decade. National Academy Press, Washington, DC.

NAS (National Academy of Sciences), 1993: Science, technology, and the Federal Government: National goals for a new era. Committee on Science, Engineering, and Public Policy. National Academy Press, Washington, DC.

Naunton, E., 1992: Warning: Andrew's bruises will linger. *The Miami Herald*, September 6, 2J.

Neumann, C.J., 1993: Global overview. In: *Global Guide to Tropical Cyclone Forecasting*. World Meteorological Organization Technical Document, WMO/TD No. 560, Tropical Cyclone Programme, Report No. TCP-31, Geneva, Switzerland, Chapter 1.

Neumann, C.J., 1996: Comments on "Determining cyclone frequencies using equal-area circles". *Mon. Wea. Rev.*, **124**, 1044–1045.

Neumann, C.J., B.R. Jarvinen, C.J. McAdie and J.D. Elms, 1993: Tropical cyclones of the North Atlantic Ocean, 1871–1992. Fourth Revision, NOAA Historical Climatology Series 6–2, Prepared by the National Climatic Data Center, Asheville, NC in cooperation with the National Hurricane Center, Coral Gables, FL, 193 pp.

Nicholls, M.E. and R.A. Pielke, 1995: A numerical investigation of the effect of vertical wind shear on tropical cyclone intensification. *21st Conference on Hurricanes and Tropical Meteorology*, AMS, Boston, 24–28 April, Miami, FL, 339–341.

Nilson, D., 1985: Conclusion: Natural hazard political contexts and adoption strategies. *Policy Studies Rev.*, **4**, 734–738.

NOAA (National Oceanic and Atmospheric Administration), 1993: "Hurricane!" A Familiarization Booklet. #NOAA PA 91001. NWS, Silver Springs, MD, 36 pp.

NOAA (National Oceanic and Atmospheric Administration), 1994: The great flood of 1993. Natural Disaster Survey Report, February, US Department of Commerce.

Noonan, B., 1993: Catastrophes: The new math. *Best's Review*, August, 41–44, 83.

NOVA, 1993: "Hurricane." Television program produced by the Public Broadcasting System.

270   HURRICANES: THEIR NATURE AND IMPACT ON SOCIETY

Novlan, D.J. and W.M. Gray, 1974: Hurricane-spawned tornadoes. *Mon. Wea. Rev.*, 102, 476–488.
NRC (National Research Council), 1989: Opportunities to improve marine forecasting. National Academy Press, Washington, DC.
NYT (*New York Times*), 1994: In Florida storm hits crops hard. *The New York Times*, November 20, A10.
O'Brien, J.J., T.S. Richards and A.C. Davis, 1996: The effects of El Niño on U.S. landfalling hurricanes. *Bull. Am. Meteor. Soc.*, 77, 773–774.
Olson, R.S. and D.C. Nilson, 1982: Public policy analysis and hazards research: Natural complements. *Soc. Sci. J.*, 19, 89–103.
Ooyama, K., 1969: Numerical simulation of the life-cycle of tropical cyclones. *J. Atmos. Sci.*, 26, 3–40.
Palm, R., 1990: *Natural Hazards: An Integrative Framework for Research and Planning*, Johns Hopkins University Press, Baltimore, MD.
Palmén, E.H., 1948: On the formation and structure of tropical cyclones. *Geophysica*, 3, 26–38.
Pasch, R.J. and L.A. Avila, 1994: Atlantic tropical systems of 1992. *Mon. Wea. Rev.*, 122, 539–548.
Penn, I. and C. Evans, 1992: Dade residents swamp Broward, seeking supplies. *The Miami Herald*, August 26, 1A.
Petrosyants, M.A. and E.K. Semenov, 1995: Individual potential of tropical cyclone genesis. *Atmos. Oceanic Phys.*, 31, 327–335.
Pfeffer, R.L. and M. Challa, 1992: The role of environmental asymmetries in Atlantic hurricane formation. *J. Atmos. Sci.*, 49, 1051–1059.
Pielke, Jr, R.A., 1994: Scientific information and global change policymaking. *Climate Change*, 28, 315–319.
Pielke, Jr., R.A., 1997: Reframing the U.S. hurricane problem. *Soc. Natural Resources*, 10(5), 485–499.
Pielke, Sr, R.A., 1990: *The Hurricane*, Routledge Press, London, 228 pp.
Pielke, Jr, R.A. and M.H. Glantz, 1995: Serving science and society: Lessons from large-scale atmospheric science programs. *Bull. Am. Meteor. Soc.*, 76, 2445–2458.
Pielke, Jr, R.A. and C. Landsea, 1997: Trends in U.S. hurricane losses, 1925–1995. 22nd Conference on Hurricanes and Tropical Meteorology, 19–23 May, 1997, Fort Collins, CO.
Pielke, Jr, R.A., J. Kimpel, C. Adams, J. Baker, S. Changnon, K. Heideman, P. Leavitt, R.N. Keener, J. McCarthy, K. Miller, A. Murphy, R.S. Pulwarty, R. Roth, E.M. Stanley Sr, T. Stewart, and T. Zacharias, 1997: Societal aspects of weather: Report of the Sixth Prospectus Development Team of the U.S. Weather Research Program to NOAA and NSF. *Bull. Am. Meteor. Soc.*, 78(5), 000–000.
Pielke, Jr, R.A. and N. Waage, 1987: Note on a definition of normal weather. *National Weather Digest*, 12, 20–22.
Pierce, C.H., 1939: The meteorological history of the New England hurricane of Sept. 21, 1938. *Mon. Wea. Rev.*, 67, 237–285.
Pitts, L. 1992: After Andrew – scared, scarred, . . . and blessed. *Miami Herald*, August 26th, 18a.
Powell, M.D., 1982: The transition of the Hurricane Frederic boundary-layer wind field from the open Gulf of Mexico to landfall. *Mon. Wea. Rev.*, 110, 1912–1932.
Powell, M.D. and S.H. Houston, 1996: Hurricane Andrew's landfall in south Florida. Part II: Surface wind fields and potential real-time applications. *Wea. Forecasting*, 11, 329–349.
Powell, M.D., S.H. Houston and T.A. Reinhold, 1996: Hurricane Andrew's landfall in

south Florida. Part I: Standardizing measurements for documentation of surface wind fields. *Wea. Forecasting*, **11**, 304–328.

Public Administration Review, 1985: Emergency management: A challenge for public administration. *Public Administration Review*, Special issue, Vol. 45, January.

Puri, K. and G.J. Holland, 1993: Numerical track prediction models. In: World Meteorological Organization Technical Document, WMO/TD- No. 560, Report No. TCP-31, Global Guide to Tropical Cyclone Forecasting, Chapter 8.

Ramage, C.S., 1959: Hurricane development. *J. Meteor.*, **16**, 227–237.

Rappaport, E., 1993: Preliminary report (updated 2 March 1993) Hurricane Andrew 16–28 August 1992. National Hurricane Center, Miami, FL.

Rappaport, E.N. and J. Fernandez-Partagas, 1995: The deadliest Atlantic tropical cyclones, 1492–1994. NOAA/NWS Technical Memorandum NWS NHC-47 (January). Coral Gables, FL, 41 pp.

Rasmussen, E., 1985: A polar low development over the Barents Sea. *Tellus*, **37A**, 407–418.

Rein, M. and S.H. White, 1977: Policy research: Belief and doubt. *Policy Anal.*, **3**, 239–271.

Rensberger, B., 1995: The cost of curiosity. *The Washington Post Weekly Edition*, January 16–22, 11–12.

Reuter's News Service, 1996: Modern warnings system cuts hurricane deaths in U.S. Reuter's News Service, 18 January.

Riehl, H., 1948: On the formation of typhoons. *J. Meteor.*, **5**, 247–264.

Riehl, H., 1954: *Tropical Meteorology*, McGraw Hill Book Company, New York, 392 pp. 867–876.

Riehl, H. and N.M. Burgner, 1950: Further studies of the movement and formation of hurricanes and their forecasting. *Bull. Am. Meteor. Soc.*, **1**, 244–253.

Riehl, H. and R.J. Shafer, 1946: The recurvature of tropical storms *J. Meteor.*, **1**, 42–54.

Rippey, B., 1997: Weatherwatch – August 1996. *Weatherwise*, **49**, 51–53.

Rodriguez, H., 1997: A socioeconomic analysis of hurricanes in Puerto Rico: An overview of disaster mitigation and preparedness, pp. 121–146. In H.F. Diaz and R. Pulwarty, Eds, *Hurricanes: Climate and Socioeconomic Impacts*, Springer, Heidelburg, Germany.

Rosenfeld, J., 1996: Cities built on sand. *Weatherwise*, **49**, 20–27.

Roth, R.R., Sr, 1996: The property casualty insurance industry and the weather of 1991–1994. In: *Impacts and Responses of the Weather Insurance Industry to Recent Weather Extremes*, Final Report to the University Corporation for Atmospheric Research, Changnon Climatologist, Mahomet, IL, 101–132.

RPI, 1997: Tropical cyclones and climate variability: A research agenda for the next century. Recommendations from the RPI Workshop on Tropical Cyclone Prediction and Climate Variability, 12–14 June, 1996.

Sadler, J.C., 1976: A role of the tropical upper tropospheric trough in early season typhoon development. *Mon. Wea. Rev.*, **104**, 1266–1278.

Salmon, J. and D. Henningson, 1987: Prior planning for post-hurricane Reconstruction. Report Number 88 (January), Florida Sea Grant College Program, University of Florida, Gainesville, FL.

Sawyer, K., 1995: Predicting a hurricane's twists and turns: Better computers help forecasters make more accurate calls. *Washington Post National Weekly Edition*, August 7–13, 34.

SCBHUA (Senate Committee on Banking, Housing, and Urban Affairs), 1992: Hearing on The National Flood Insurance Reform Act of 1992, Statement of Dr.

Robert Sheets before the Subcommittee on Housing and Urban Affairs, 27 July, 102–913.

Schware, F.K., 1970: The unprecedented rain associated with the remnants of Hurricane Camille. *Mon. Wea. Rev.*, **98**, 851–859.

Schwerdt, R.W., F.P. Ho and R.R. Watkins, 1979: Meteorological criteria for standard project hurricane and probable maximum hurricane windfields, Gulf and East Coasts of the United States. NOAA Technical Report NWS 23, U.S. Department of Commerce, National Oceanic and Atmospheric Administration, National Weather Service, Washington, DC, September 1979.

Scism, L., 1996: Nationwide to unveil rules to cut risks after hurricanes. *Wall Street Journal*, October 10, 1996.

Seabrook, C., 1990: Hugo's lessons: Gone with the wind. *Atlanta Constitution*, April 17, B6.

Shabman, L.S., 1994: Responding to the 1993 flood: The restoration option. *Water Resources Update*, **95**, 26–30.

Shapiro, L.J., 1989: The relationship of the quasi-biennial oscillation to Atlantic tropical storm activity. *Mon. Wea. Rev.*, **117**, 1545–1552.

Sheets, R.C., 1990: The National Hurricane Center – past, present, and future. *Wea. Forecasting*, **5**, 185–232.

Sheets, R., 1992: Statement before Hearing of the Senate Committee on Banking, Housing, and Urban Affairs, 1992. "The National Flood Insurance Reform Act of 1992," before the Subcommittee on Housing and Urban Affairs, 27 July, 102–913.

Sheets, R., 1993: Statement before Hearing of the House Committee on Science, Space, and Technology, 1993. "NOAA's Response to Weather Hazards – Has Nature Gone Mad?" Hearing before the Subcommittee on Space, 14 September, 103–55.

Sheets, R., 1994: Statement before Hearing of the House Committee on Public Works and Transportation, "H.R. 2873, The Natural Disaster Prevention Act of 1993," Subcommittee on Water Resources and Environment, February 2, 13 pp.

Sheets, R.C., 1995: Stormy weather. *Forum for Applied Research and Public Policy*, **10**, 5–15.

Silva, M. and T. Nickens, 1992: Dade seeks to keep tax windfall. *The Miami Herald*, October 29, 1A.

Simpson, R.H., 1946: On the movement of tropical cyclones. *Trans. Am. Geophys. Union*, **27**, 641–655.

Simpson, R.H., 1954: Hurricanes. *Scientific American*, June.

Simpson, R.H., 1978: Hurricane prediction. In: National Academy of Sciences, Geophysical Predictions. National Academy Press, Washington, DC.

Simpson, R.H., 1980: Implementation of the National Hurricane Research Project 1955–1956. Paper prepared for the 13th Technical Conference on Hurricanes and Tropical Meteorology, December 1–5, Miami Beach, FL.

Simpson, R.H. and M.B. Lawrence, 1971: Atlantic hurricane frequencies along the U.S. coastline. NOAA Technical Memorandum NWS SR-58. NWS, Silver Springs, MD.

Simpson, R.H. and H. Riehl, 1981: *The Hurricane and its Impact*, Louisiana State University Press, Baton Rouge, 398 pp.

Simpson, R.H. et al., 1978: *TYMOD: Typhoon Moderation*, Final Report prepared for the Government of the Philippines, Virginia Technology, Arlington.

Sims, J.H. and D.D. Baumann, 1983: Educational programs and human response to natural hazards. *Environ. Behav.*, **15**, 165–189.

Slevin, P. and R. Greene, 1992: A deepening crisis downpour adds to misery of Andrew's victims. *The Miami Herald*, August 30, 1A.

Slovic, P., 1993: Perceived risk, trust, and democracy. *Risk Anal.*, **13**, 675–682.

Smith, K., 1992: *Environmental Hazards: Assessing Risks and Reducing Disaster*, Routledge, London.

Sorensen, J.H., 1993: Warning systems and the public warning response. Workshop on Socioeconomic Aspects of Disaster in Central America, San Jose, Costa Rica, 21–23 January.

Sorensen, J.H. and D.S. Mileti, 1988: Warning and evacuation: Answering some basic questions. *Indus. Case Quart.*, **5**, 33–61.

Southern, R.L., 1992: Savage impact of recent catastrophic tropical cyclones emphasizes urgent need to enhance warning/response and mitigation systems in the Asia/Pacific region (unpublished).

Stern, P.C., 1986: Blind spots in policy analysis: What economics doesn't say about energy use. *J. Policy Anal. Manage.*, **5**, 200–227.

Stewart, T.R., 1997: Forecast value: Descriptive studies. In: *Value of Weather and Climate Forecasts*, R.W. Katz and A.H. Murphy, Eds, Cambridge University Press, Cambridge, UK.

STFDR (Senate Task Force on Disaster Relief), 1995: Document 104–4, March 15. GPO, Washington, DC.

Stone, G.W., J.M. Grymes III, C.K. Armbruster, J.P. Xu and O.K. Huh, 1996: Researchers study impact of Hurricane Opal on Florida coast. *Eos*, **77**, 181, 184.

Strouse, C., 1992: Tiny rise in assessments = lean budgets. *The Miami Herald*, June 18, 1B.

Sugg, A.L., 1967: Economic aspects of hurricanes. *Mon. Wea. Rev.*, **95**, 143–146.

Swarns, R.L., 1992: Uninsured left with less than nothing: Devastated homes, monumental debts. *The Miami Herald*, November 2, 1A.

Swenson, E.L., 1993: Testimony for the U.S. House of Representatives, Committee on Veterans Affairs, 103–4. GPO, Washington, DC.

Sylves, R., 1996: The political process of presidential disaster declarations. Quick Response Report #86, University of Colorado, Natural Hazards Research and Applications Information Center, http://adder.colorado.edu/hazctr/qr/qr86.html

Tanfani, J., 1992: Miami tallies storm cost, may end up with net gain. *The Miami Herald*, September 11, 6B.

Tannehill, I.R., 1952: *Hurricanes: Their Nature and History*, 8th edition, Princeton University Press, Princeton, NJ.

Tennekes, H., 1990: A sideways look at climate research. *Weather*, **45**, 67–68.

Texas Coastal and Marine Council, 1974: *Hurricane Awareness Briefings*, Texas Coastal and Marine Council, Austin, TX.

TFFFCP (Task Force on Federal Flood Control Policy), 1966: *A Unified National Program for Managing Flood Losses*, Report No. 67–663, United States Government Printing Office, Washington, DC.

Thorne, N.J., 1984: Reinsurance and natural disasters. *The Geneva Papers on Risk and Insurance*, **9**, 167–174.

Timmerman, P., 1981: Vulnerability, resilience and the collapse of society: A review of models and possible climatic applications. Environmental Monograph 1, Institute for Environmental Studies, University of Toronto, Toronto, Canada.

Torgerson, D., 1985: Contextual orientation in policy analysis: The contribution of Harold D. Lasswell. *Policy Sci.*, **18**, 241–261.

Trausch, S., 1992: Hurricane profiteers have we got a great deal for you! *The Miami Herald*, September 8, 21A.

UNEP, 1989: Criteria for assessing vulnerability to sea-level rise: A global inventory to high risk areas. UNEP, Nairobi.

Ungar, S., 1995: Social scares and global warming: Beyond the Rio convention. *Soc. Nat. Resources*, **8**, 443–456.

USACE/FEMA, 1993: Hurricane Andrew assessment: Review of hurricane evacuation studies utilization and information dissemination, Florida Version. Prepared by Post, Buckley, Schuhe, Jernigan, Inc., Tallahassee, FL.
USDI/NPS (US Department of the Interior, National Park Service), 1994. Hurricane Andrew, 1992: The National Park Service response in South Florida (August). Washington, DC: Department of the Interior.
USWRP, 1994: United States Weather Research Program: Implementation Plan. Congressional Submission, Committee on Earth and Environmental Sciences, January 15.
Van Natta, D. 1992: Precautions, warnings cut storm death toll but luck also played a big role. *The Miami Herald*, September 14, 13A.
Viglucci, A., 1992: Victims to FEMA: We've been ignored. *The Miami Herald*, September 8, 1B.
Wakimoto, R.M. and P.G. Black, 1994: Damage survey of Hurricane Andrew and its relationship to the eyewall. *Bull. Am. Meteor. Soc.*, **75**, 189–202.
Wamsted, D., 1993: Insurance industry wakes up to threat of global climate change. *Environ. Week*, **6**, October 7.
Wang, Y. and G.J. Holland, 1996: The beta drift of baroclinic vortices. Part II: Diabatic vortices. *J. Atmos. Sci.*, **53**, 3737–3756.
West, C. and D. Lenze, 1994: Modeling the regional impact of natural disaster and recover: A general framework and an application of Hurricane Andrew. *Int. Regional Sci. Rev.*, **17**, 121–150.
White, G.F., 1994: A perspective on reducing losses from natural hazards. *Bull. Am. Meteor. Soc.*, **75**, 1237–1240.
White, G.F. and J.E. Haas, 1975: *Assessment of Research on Natural Hazards*, The MIT Press, Cambridge, MA, 487 pp.
Whoriskey, P., 1992: U.S. experts fault code on mobile home safety. *The Miami Herald*, September 10, 1A.
Wildavsky, A., 1979: *Speaking Truth to Power: The Art and Craft of Policy Analysis*, Little, Brown, Boston, MA.
Williams, J., 1992: *The Weather Book*. New York: Vintage Books.
Willoughby, H.E., 1979: *Some Aspects of the Dynamics in Hurricane Anita of 1977*, Coral Gables, FL, NOAA Technical Memorandum, ERL NHEML-5.
Willoughby, H.E., 1990: Temporal changes in the primary circulation in tropical cyclones. *J. Atmos. Sci.*, **47**, 242–264.
Willoughby, H.E. and P.G. Black, 1996: Hurricane Andrew in Florida: Dynamics of a disaster. *Bull. Am. Meteor. Soc.*, **77**, 543–549.
Willoughby, H.E., J.A. Clos and M.G. Shoreibah, 1982: Concentric eye walls, secondary wind maxima, and the evolution of the hurricane vortex. *J. Atmos. Sci.*, **41**, 1169–1186.
Willoughby, H.E., J.M. Masters and C.W. Landsea, 1989: A record minimum sea level pressure observed in Hurricane Gilbert. *Mon. Wea. Rev.*, **117**, 2824–2828.
Willoughby, H.E., D.P. Jorgensen, R.A. Black and S.L. Rosenthal, 1985: Project STORMFURY: A scientific chronicle 1962–1983. *Bull. Am. Meteor. Soc.*, **66**, 505–514.
Wilson, N.C., 1994: Surge of hurricanes and floods perturbs insurance industry. *J. Meteor.*, **19**, 3–7.
Winchester, P., 1991: *Power, Choice and Vulnerability: A Case Study in Disaster Mismanagement in South India, 1977–78*, James & James.
Winkler, R.L. and A.H. Murphy, 1985: Decision analysis. In: *Probability, Statistics, and Decisionmaking in the Atmospheric Sciences*, A.H. Murphy and R.W. Katz, Eds, Westview Press, Boulder, CO, 493–524.

Wolensky, R.P. and K.C. Wolensky, 1990: Local government's problem with disaster management: A literature review and structural analysis. *Policy Studies Rev.*, **9**, 703–725.

World Meteorological Organization, 1993: *Global guide to tropical cyclone forecasting.* WMO/TD – NO. 560, Report No. TCP-31, G. Holland, Ed.

Wu, C.-C. and Y. Kurihara, 1996: A numerical study of the feedback mechanisms of hurricane-environment interaction on hurricane movement from the potential vorticity perspective. *J. Atmos. Sci.*, **53**, 2264–2282.

Xie, L. and L.J. Pietrafesa, 1995: Modeling the flooding around the Pamlico-Albemarle sounds caused by hurricanes, Part II: Predicting the flooding caused by Emily. *Preprints, 21st Conference on Hurricanes and Tropical Meteorology*, Miami, FL, American Meteorological Society, 445–447.

Yarnall, B., 1994: Agricultural decollectivization and vulnerability to environmental change. *Global Environ. Change*, **4**, 229–243.

Index

additional readings 194–7
aerodynamic roughness 81
Alicia 20, 24
Andrew
  damages and casualties 163–70
  forecast 3–6, 156
  evacuation 158–62
  track and intensity 156–7
  impact 7–9, 163
    building codes 170–6
    direct damages 163–9
    insurance 176–7
    relation between central pressure and
      wind speed 80
    response 10–13, 177
    recovery 177–8
    restoration 179–80
assessment of vulnerability 59
  incidence assessment 65
  societal vulnerability to tropical
    cyclones 64–5
  defined 64
  tropical cyclone risk assessment
    59–64
atmospheric pressure 76
  defined 68

building codes 143, 170–6

Camille 3, 19–20, 23–4, 39, 49
catastrophe models 59–62
centrifugal force
  defined 72
climate
  defined 139–41, 185–6
climate change 47–8, 185–6
community planning 145
condensation 71

convergence
  defined 69
  illustrated 70
coriolis effect
  defined 68, 100
  illustrated 69–70
criteria for development and
  intensification 87–90

damage estimation 134–6
damages in US (normalized, 1925–1995)
  198
deaths and damages, US (1900–1995)
  202
  30 deadliest, US (1900–1995) 199
  30 costliest, US (1900–1995)
    200–1
decay 79–81
decision-maker
  defined 141
descriptive assessments 115–17
divergence
  defined 69
  illustrated 70
deposition 71

El Niño-Southern Oscillation (ENSO)
  109
evacuation 141, 158–62
evaporation of water 81
exposure assessment 65
extratropical cyclone
  defined 15
extreme weather events 25–6, 140–1
eye
  definition 72
  illustration of 74, 78
  satellite images of 75
  size 88

eye wall
  concentric eye wall cycles 73
  defined 72
  double eye wall 73
  illustration 78
  radius of maximum winds 74

feeder bands
  defined 72
FEMA 58, 113, 162, 168, 171
forecasts 92
  average error 107
  example of Hurricane Hugo 106
  interaction of steering current and
      hurricane 99–101
  internal flow 102
  movement 92
  steering current 92–8
    defined 93
    illustrated 93
  track, intensity, and seasonal
      forecasting 84, 94–8, 101–3,
      219
    attempts at modification 111–12
    intensity change predictions 106–8
    seasonal predictions 109–11
    track predictions 102–5
    value to society 113–17
forecast value 113–17
Frederic 20, 23–4
frequency
  Atlantic Basin (1950–1996) 86, 89
  world-wide 218
fundamentals 181
  knowledge to action 181
  ten important lessons 182–90

Galveston (1900) 19
geographic and seasonal distribution 68,
    85
  global illustration 84
  in the Atlantic Ocean Basin 87–90
  movement 85–6
  origin 85
giant hurricanes 88
global tropical cyclones tracks
  illustration of 84
Gray, William 40–1, 109–11, 184
guide for local hurricane decision makers
    233–55
Gulfstream IV 109

Hazel 20, 22–3
high shear environment 79
  illustrated 80
Hugo 21, 24
hurricane
  defined 15
  in the past century 19–24
  North American history 16–18
hypercanes
  defined 78

impacts 118
  ocean 118
  land impacts at the Coast and a Short
      Distance Inland 119
  inland impacts 127–30
  rainfall
    forecast 129
    observed 120, 125–6, 130, 132
  storm surge 119–20
  storm surge analysis 121
  tornadoes 120, 125–6
    observed 127–8
  winds 119, 122–4
  societal impacts 131–3
    estimating damages 134–6
      aggregation: the problem of
          benefits 136
      attribution: the problem of
          causation 135
      comparison: the problem of
          demographic change 137
      contingency 134–5
      quantification: the problem with
          measurement, 135–6
insurance 145–7, 176–7
intensification 80, 87–90
intensity
  defined 39–40
  predicting 39–40

June solstice
  defined 87

Labor Day, Florida Keys Storm of 1935
    77
landfall
  frequency 43–6
  impacts 43–6, 118–38
land use 143–4
largest radius of tropical storm winds 87
latent heating 71

life of a tropical cyclone 68
  birth and growth 68
  criteria for development and
    intensification 87–90
  cyclonic circulation 82
  high pressure 82
  decay 79–81
  maturity 68–71

major hurricane landfalls on US mainland
  coast (1901–1996) 206–8
maximum gusts 124
maximum potential intensity 78
mean direction of motion of tropical
  cyclones world-wide 84, 219
MEOW 121
midget hurricanes 88
models
  BAM 105
  CLIPER 102
  GFDL 105
  MRF 105
  NHC 90 102
  UK 90 105
  VICBAR 105
model simulations
  eye and eye wall 78
  horizontal cross section 79
  hurricane wind circulation 76
  storm surge 123
  track 101
modification 111–13
movement 85–6, 92

names
  Arafura Sea and Gulf of Carpentaria
    211
  Atlantic 18, 209
  Eastern Australia Region 214
  Eastern North Pacific 210
  Fiji Region 215
  Northern Australia Region 212
  Solomon Sea and Gulf of Papua 211
  Southeast Indian Ocean 217
  Southwest Indian Ocean 216
  Western Australia Region 213
  Western North Pacific Ocean 217
National Hurricane Center 102
National Oceanographic and
    Atmospheric Administration
    (NOAA) 109
New England (1938) 22

normal to large hurricanes 88
normal weather event 140–1
number of tropical cyclones and
  hurricanes (1950–1996) 203

observations of
  hurricane structure 77
  surface features 119
oscillation in movement 102
outflow jets
  defined 100

P-3 Orions 109
polar lows
  defined 85
policymaker
  defined 141
population at risk 49–54
preparedness 56–9
prescriptive assessments 115–17
pressure systems
  Azores High 87
  Bermuda High 87
  defined 68
Project Stormfury 112–13
property at risk 54–6

rainfall
  forecast 129
  observed 120, 125–6, 130, 132
  rapid acceleration 99
  relation between wind speed and
    destructive force 124
retired names 20–1

Saffir-Simpson Damage-Potential scale
  defined 16–17
  illustrated 90
satellite images
  infrared 73
  visible 75
science and policy problems
  as an extreme meteorological event 15
  extreme weather events 25
  science in service to society 26–9
science policy 14, 26–30
seasonal predictions 109–11
severe cyclonic storms 85
SHIFOR
  defined 108
SHIPS
  defined 108

silver iodide 112
Simpson, Robert 15–16, 22–3, 148
sizes of tropical cyclones 87–8
SLOSH basins 116, 121
  Charleston, South Carolina 123
  US Gulf and Atlantic coasts 122
SLOSH modeling 57–8, 121–2
smallest radius of tropical storm winds
  87
societal responses 139–40
  long-term social and decision
    processes 141
  preparing for evacuation 141
  preparing for impacts 142–3
  preparing for recovery 144–7
  short-term decision processes 148
  hurricane track prediction
    148–51
  recovery and restoration 154–5
  surviving the storm 152
special cases 83–4
  cold-core lows 83
  subtropical cyclones 83
  tropical cyclone genesis 83
  tropical low pressure systems 83
spiral bands 72
steering current 92–101
  defined 93
  illustrated 93
storm surge 119–20
  analysis 121
  during Andrew 160–2
  modeling 123
stratosphere
  defined 69
stratospheric quasibiennial oscillation
  defined 109
supercooled water
  defined 112

TDL 121
thermal heat engine 71
tornado 125–7
tornado swarms 128
track models 101
trade winds
  defined 87

tropical cyclone
  defined 15
tropical depression
  defined 15, 71
tropical disturbance
  defined 15, 71
tropical low
  defined 15
tropical storm 72
  defined 15
tropical waves
  defined 87
troposphere
  defined 87
typhoon
  defined 15, 85

units of wind speed, distance, and
  temperature 256–7
US hurricane problem 31
  conventional framing of problem
    35–6
  definition 31
  societal vulnerability as an alternative
    framing; 37

vulnerability 38
  assessment 59–64, 191–2
  climate change 47–8
  exposure 49–58
  framework of xv–xvi, 64–7
  hurricane incidence 39–46

warning responsibility areas for tropical
  cyclones
  Atlantic and Eastern North Pacific
    104
  world-wide 220–9
  watch and warning dissemination
    151
weather
  defined 139–41
  winds 119, 122–4
wind shear 71
  effect on intensification 80
winter storms 119
WSR-D-88 system 130